Railway Engineering, Systems, and Safety

Organizing Committee Members

R Aylward (Chair)
Eversholt Train Leasing Company

Alan Constable
London Underground Limited

D Gillan
Railway Industry Association

Roger Goodall
Loughborough University of Technology

Richard Gostling
British Rail Research

John Harris
ADTranz

Phil Kay
London Underground Limited

Colin May
Maresfield

Tony May
Centre Exhibitions, National Exhibition Centre

Pat O'Connor
Metro Consulting

Paul Ramsey
BR Infrastructure Services HQ

Mike Richards
Centre Exhibitions, National Exhibition Centre

Paul Robinson
RFS (E) Limited

Dave Russell
SAB WABCO (Bromborough) Limited

Dave Saunders
Network Train Engineering Services

Jon Seddon
Interfleet Technology

Graeme Smart
ABB British Wheelsets Ltd

Brian Tallis
Severn Advertising Limited

Keith Ware
York

Brian West
Brush Traction

Siep Wijsenbeek
Design Triangle

IMechE
Seminar Publication

150th Anniversary
1847 – 1997

Railway Engineering, Systems, and Safety

Selected papers from Railtech 96

Organized jointly by the Railway Division of the Institution of Mechanical Engineers
and Centre Exhibitions, a member of the NEC Group

IMechE Seminar Publication 1996–16

Published by Mechanical Engineering Publications Limited for
The Institution of Mechanical Engineers, Bury St Edmunds and London.

First Published 1996

ISSN 1357–9193
ISBN 1 86058 015 7
ISBN 978-1-86058-015-4

A CIP catalogue record for this book is available from the British Library.

Contents

Related Titles of Interest

Title	Author	ISBN
Railway Rolling Stock	IMechE Seminar 1996–17	1 86058 016 5
Design, Reliability, and Maintenance for Rail Transport	IMechE Seminar 1996–18	1 86058 017 3
Rail Traction and Braking	IMechE Seminar 1996–19	1 86058 018 1
Set of 4 Railtech Volumes (including the 3 titles listed above and this title)	IMechE Seminar 1996	1 86058 050 5
Implementing Rail Projects	IMechE Conference 1995–1	0 85298 948 2
Railways for Tomorrow's Passengers	IMechE Conference 1993–7	0 85298 859 1

For the full range of titles published by MEP contact:

Sales Department
Mechanical Engineering Publications Limited
Northgate Avenue
Bury St Edmunds
Suffolk
IP32 6BW
England

Tel: 01284 763277
Fax: 01284 704006

C511/7/019/96

Platform edge doors – a first for London Underground

M A HEARN BEng, AMIEE and **A T NORMAN** MIMechE, FIEE, FIMinE
London Underground Limited, Jubilee Line Extension Project, London, UK
M O'BRIEN AMSaRs, MIDiagE, MISM
Westinghouse Brakes Limited, Wiltshire, UK

SYNOPSIS

London Underground Limited, for the first time in its history is installing Platform Edge Doors (PED's) as part of the new Jubilee Line Extension Project.

These doors will introduce a radical new concept to British Metro Systems. Great interest has been generated by several parties during the development of this Project and this paper will provide an overview from the perspective of both the Client and the Contractor.

1.   Introduction

London Underground is currently implementing its plan to extend the existing Jubilee Line to parts of South and East London servicing Docklands. This will be the biggest addition to the underground in over twenty years.

The Project (Jubilee Line Extension Project) will bring about a major improvement to London's public transport systems, particularly along an axis from the West End through the South Bank to the developing areas of Docklands and inner East London. The end product will be a modern mass transit railway 37km long of which 16km will be new.

Fifty nine new trains will operate on the extended Jubilee Line. Utilising the latest proven technology the service will provide up to 27 trains per hour with typical journey times between stations being of order 2 mins.

Platform Edge Doors (PED's) will be installed for the first time on the Underground at each of JLE's eight new sub-surface stations.

Each of the seventeen new underground station platforms from Westminster to North Greenwich will be equipped with a platform edge screen in the form of a glazed barrier along the edge of the platform adjacent to the track, incorporating automatic sliding doors.

This paper now goes on to describe the PED concept, design development and finally the installation activities associated with the project.

The joint authorship of the paper will allow the input of both the client (London Underground) and contractor (Westinghouse Brakes Ltd.) to be incorporated.

2.0   DESIGN & MANUFACTURE

2.1   Jubilee Line Extension (JLE) Requirements

JLE requirements for Platform edge doors were documented in the form of a performance specification produced by the JLE project team. The WBL objectives were to design, manufacture and install the platform edge doors in accordance with the JLE specification. In addition the system to be produced, required high safety, reliability and maintenance standards.

The prime purpose of the PED screen is to improve the passenger environment by reducing wind speeds experienced by the passenger and to provide the passenger with a secure platform edge area. The door screen also reduces the effects of train noise.

2.2   System Concept/Description

The JLE Platform Edge Door (PED) system comprises an Architecturally styled free standing curtain wall structure attached to the platform edge providing a barrier between the passengers and the track. The height of the barrier is approximately 2.5m, just above train height, and runs the full length of the public section of each individual platform.

The PED system contains a number of automatic bi-parting doors spaced at intervals aligned with the doors of a train consisting of seven cars. The PED's comprise both Symmetrical and Asymmetrical doorways to correspond with the car door arrangement. Symmetrical Doors align with the two leaf and Asymmetrical with the single leaf doorways of the car. The platform edge screen consists of a total of 14 symmetrical, 12 asymmetrical doorways plus 2 Driver doorways (also asymmetrical). The operation of the doors for the seventh car is inhibited to allow the JLE 6 car train system to run with a view to future expansion to seven cars .

The spaces between the Automatic Doors along the structure are filled by fixed Glass Panels. These panels, in the form of frangible toughened safety glass, are designed to withstand crowd pressure loading, pressure pulses, wind loading and passenger impact loading, albeit the glass is also designed to be broken if required with a special tool, in an emergency, by operational staff.

At the extreme ends of the platform there are two non powered hinged doors sited at right angles to the structure for tunnel access. Adjacent to the doors are mounted the Headwall/Tailwall units (near to the front and rear of train respectively) these units effectively close off the public platform length. The Headwall/Tailwall units are also utilised to mount items of plant such as Tunnel telephones, signal repeater units, and several other platform related items.

The support gear and actuating equipment for the PED's consist of a door control unit providing the intelligent control interface with the 110 V DC motor and gearbox. Motion to the doors is provided through a simple belt and pulley drive system. The actuating equipment is enclosed within a locked covers (hinged for access), that run the full platform length.

The PED Structure consists of a bracket and post arrangement which is cantilevered from and 'stands off' from the concrete face of the platform on spacers. The PED support brackets and posts are secured to the vertical face of the platform using chemically bonded fixing bolts. A Header beam is bolted directly to the top of the posts forming the main structure and support for the PED system.

Platform Side Concept   Trackside Concept

2.3   System Design/Development

WBL have many years of experience in the field of door systems and have similar systems operating in Singapore and Stansted airport. In addition, WBL, have produced thousands of train doors operating around the world.

The door design was developed using WBL engineering expertise and was subject to design scrutiny including detailed Safety, Reliability and Maintainability assessments. The system was also subjected to substantial development, prototype compliance and interface testing.

The door actuating equipment has been developed over several years, development consisting of functional operation and life testing. The life testing, over 1.5 million open and close operations is equivalent to 7.5 years of cycling of a doorway on the JLE system. The drive gear utilised is similar to that used on the platform edge doors in the Singapore Metro and Stansted airport people mover.

Following contract award form JLE, WBL commenced the conceptual design. A 'dummy' platform was constructed in a building at the WBL works and two prototype JLE doorways were manufactured and installed. The Prototype doorways were functionally tested, including a 1.5 million cycle test. The results of this were recorded and fed back into the design development as system improvements.

Following completion of site prototype testing and incorporation of system design improvements WBL upgraded the design to a 'ready for production' level. Manufacture commenced on programme and the system was subjected to compliance testing in accordance with the JLE specification. Following successful compliance testing the 'ready for production' doorways were subjected to a life endurance test of 2.5 million cycles equating to 12.5 years cycling of a doorway on the JLE system.

After proving the design during the compliance testing, and about midway through the life testing, a 25 metre platform screen, equivalent to one and a quarter train cars in length, was erected at the WBL works. The system was erected to simulate and test passenger/operator and PED interfaces. The Passenger Interface Trial (PIT) was undertaken to simulate passenger movement, reactions etc. In addition it was used to simulate operational and maintenance procedures. The PIT was built using production drawings and also provided a trial run of actual main contract manufacture and installation procedures.

Passenger Interface Trial Doors

2.4 Interfaces

During design development WBL had to ensure that the PED's would fully integrate with the JLE system. Interface design therefore formed a major part of the design process, the main interfaces being described below:

Train: The train interface included train length, door alignment, door opening/closing speeds, door opening width. The PED system needs to be a set distance from the Kinematic profile of the train to ensure that train/PED clearances are maintained.

Signalling:	The signalling equipment provides the open and close commands for PED's and the signalling aerials in the form of Leaky feeder antenna are mounted directly below the PED structure.
Comms:	The PED reports faults to the JLE SCADA system allowing preventative/corrective maintenance. Several station announcements can be derived from the PED's. The PED's are also used to mount the CCTV antenna.
Civil Structure:	The Civil Interfaces relate to physical positioning of the PED's. The strength and condition of the platform to which the PED's are mounted is critical to a successful installation.
Architecture:	The PED's are designed utilising architectural input to ensure that the styling is consistent with that of each individual station.

2.5 Operation

The train approaches the platform under automatic train operation (ATO), is slowed, stopped at the correct position and aligned with the PED's. The train door open command is initiated by the train driver by pressing the "door open" buttons, the signalling system then registers the signal and confirms that the train has stopped and docked in the correct position, on confirmation of this the signalling system provides a signal for the both train doors and PED doors to open.

To leave the station, the driver initiates the "doors closed" signal by pressing the "doors closed" buttons, the signalling system issues the "doors closed" signals to the train and to the PED's. Once the train doors and PED doors are closed and locked a signal is sent from both the train and PED's and the train is allowed to depart.

The PED system can also be operated manually by specifically designed platform switches which allow trained platform staff to operate the doorways using a staff key. Platform staff can also isolate individual doors, manually operate (open/close) a complete set of platform doors or bypass a door interlock, if required, in the unlikely event of a doorway failure.

Maintenance is considered as an integral part of the design and the system is designed to minimise maintenance downtime and reduce system disruption. A condition based maintenance system is primarily used and the condition of each doorway can be investigated and analysed using the internal diagnostic devices contained within the PED system. Several status messages are displayed directly to the SCADA system such as slow door speed, high friction etc. allowing the maintenance team to easily pinpoint potential problem areas and to maintain the system in non traffic hours. Failures occuring in service are therefore minimised ensuring optimum system availability.

2.6   Safety Features

There are several safety features that assist in the safe use of the PED system, for example:

Vital interlock system - A safety interlock system used to identify the status of each doorway and to ensure that the train cannot depart until all doorways are successfully closed and locked.

Door motion profile - The door motion profile is programmable and allows the acceleration/deceleration and operational speeds of the doorways to be pre-set. The door profile is designed to ensure that injuries do not occur in the event of a passenger obstructing the doorway. In addition the doors have an obstruction detection routine as an integral part of the profile.

Other safety features include Emergency release manual opening, key opening from platform side, doorway lighting to illuminate the open doorway threshold and to provide a visual indication of the doorway status, glass marking to allow the visually impaired identify the screens.

2.7   Final Design

WBL have designed a system that achieves the JLE requirements and WBL objectives, the final system design and installation will provide the customer, i.e. the passengers, with an aesthetically pleasing, more comfortable and safer environment in which to travel.

Operationally the system design is safe and reliable, predicted to achieve more than 12.5 years before a PED failure causes a platform to be closed. The system is easy to troubleshoot and maintain, thus reducing system downtime and optimising system availability.

Architectural Platform Concept

3.0 INSTALLATION & COMMISSIONING

As can be seen from the preceeding section great emphasis has been placed upon design development, quality assured manufacture and meticulous testing to ensure the PED's are fit for the purpose.

However ultimate system performance can only be proven on site and this section of the paper goes on to describe how the JLE/WBL Quality System is carried through the installation phase to final commissioning.

3.1 Planning & Programming

PED's are only one of the 13 main Electrical and Mechanical Contracts required to furnish the JLEP (Jubilee Line Extension Project). Given that the overall period for fixed equipment installations spans some 18 months, it is essential that each contract installation phase is carefully planned and integrated. Furthermore the interrelationship and dependencies between E&M Contractors requires interface co-ordination to form an essential part of the installation programme logic.

The JLE Project has adopted several common programming tools to achieve the above objectives. On site the key plans are denoted as:

TRIP - Track Related Installation Programme
STRIP - Station Related Installation Programme

The TRIP programme is developed to ensure sufficient overall track possession time is available to each of the E&M Contractors in order for them to complete their work and to enable Test Running of the system to commence at the specified date. The programme is in the form of a time chainage plan outlining the allocation of route sections for the installation of trackwork and E&M Services.

STRIP is also a time based activity chart which establishes that sufficient overall time is available to complete the E&M Works. However it is founded upon Civil Access dates for each area of the Stations rather than track occupation.

TRIP and STRIP are developed to produce the Engineer's Co-ordinated Installation Programme with the installation activities being finally defined in each E&M Contractor's detailed Installation Programmes.

3.2 Civil Co-ordination

The PED interface with each station structure requires considerable co-ordination throughout the construction period to ensure that the platform construction criteria and PED designs are compatible. Typical interface issues which require close attention are:-

Position of Cable Entry Ducts into the Headwall and Tailwall Units

Platform Wall Design Criteria in Terms of Strength and Position

Platform Wall Construction Tolerances

As the civil construction progresses, the E&M and Civil Interfacing continues to develop :-

- agreed access routes for the delivery of materials and men to the worksite

- attendances such as facilities for site accommodation, lay down areas, etc.

Site Induction and Safety Training of staff are also introduced at this stage.

3.3 Installation Sequence

Having co-ordinated the works with other Contractors and defined the installation plan in terms of time and space, the actual installation sequence can commence.

It is essential that works proceed smoothly from one site to the next progressing steadily from the east towards the later finished west end stations. To achieve this detailed activity sequences have been developed. Along with 'flat pack' delivery techniques and customised installation tools the whole installation proceeds on a 'production line' basis.

A typical station installation follows the following sequence:

Surveying of the Platform Area: confirming datums and marking of fundamental PED positions along the platform.

Drilling of Platform Fixing Holes - Using a series of portable custom built jigs running along the permanent track rails. Each pair of holes is gauged from the previous pair of holes by using colour coded tie bars for the appropriate distances.

Mounting of Brackets and Posts - The PED structure must accommodate the tolerances of both the platform wall and the trackwork installation.

The vertical position of the holes is determined by the drilling jig, whilst the longitudinal position is determined by using a gauge from the jig. Typically six spacers per bracket are gauged in this manner, with the intermediate spacers set from a straight edge. The brackets, weighing typically (147kg) are lowered from the platform side, again by custom built portable equipment.

Mounting of Header Beams - completes the basic structure being fitted to upper PED framework.

Laying of Floor Tiles and Screeding - proceeds after the main PED Structure is in place. This defines the finished Floor Level and the Civil Contractor is thus provided with a datum to complete the floor laying activities.

8

Installation of Electrical/Mechanical Equipment - the PED components are hand portable and can be wired and fitted out on site.

Installation of Sliding Doors, Glass Panels and Header Covers - finalises the installation phase. This is heavy work manipulating and fitting the sensitive and architecturally finished items as they are bought onto site.

3.4 Commissioning

Clearly commissioning of the PED systems requires careful interface management with several other Contractors such as Signalling and Communications.

The commissioning philosophy follows three distinct phases:-

- Pre commissioning tests
- Integrated Tests
- Test Running

The signalling system and the correct operation of all items such as the PED's from it are tested during the Test Running phases of the Project.

As part of the PED reliability tests, the doors are cycled continuously for 28 days prior to their integration into the Jubilee Line Extension Project signalling system.

Finally the whole of the railway system is tested during the Trial Running phase.

4.0 CONCLUSION

This paper is only intended to provide an overview of the Platform Edge Door systems being introduced to LUL on the Jubilee Line Extension Project.

Clearly for the system to deal with the envisaged throughput of passengers the design and ultimately the door units themselves will have to perform consistently to the highest performance and safety standards. Hopefully this paper has demonstrated the commitment of both contractor and client to produce a fully integrated door system which has set quality standards to meet those requirements.

Acknowledgement is paid to the Project Chief Engineer for giving permission to publish this paper. Credit must also go to Engineers within the JLE and WBL teams who provided information and assisted the authors.

It should be noted that any opinions expressed are those of the authors and not necessarily those of London Underground or Westinghouse Brakes Ltd.

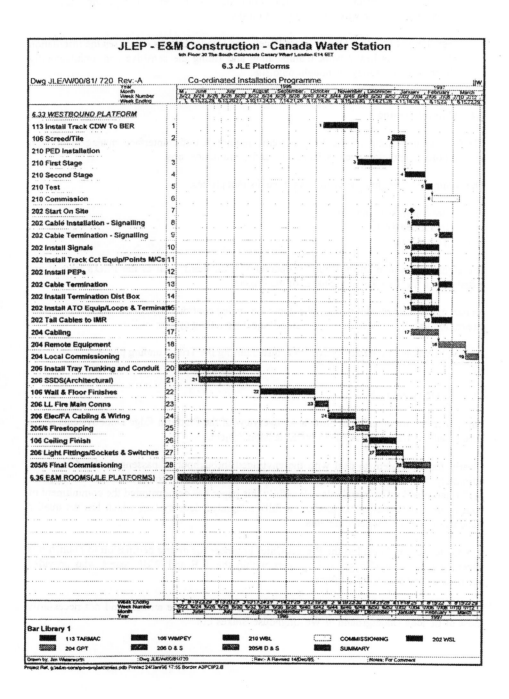

JLEP - E&M Construction - Canada Water Station
9th Floor 30 The South Colonnade Canary Wharf London E14 5ET

6.3 JLE Platforms

Dwg JLE/W/00/81/ 720 Rev:-A   Co-ordinated Installation Programme   JIW

6.33 WESTBOUND PLATFORM

Task	No.
113 Install Track CDW To BER	1
106 Screed/Tile	2
210 PED Installation	
210 First Stage	3
210 Second Stage	4
210 Test	5
210 Commission	6
202 Start On Site	7
202 Cable Installation - Signalling	8
202 Cable Termination - Signalling	9
202 Install Signals	10
202 Install Track Cct Equip/Points M/Cs	11
202 Install PEPs	12
202 Cable Termination	13
202 Install Termination Dist Box	14
202 Install ATO Equip/Loops & Terminate	15
202 Tail Cables to IMR	16
204 Cabling	17
204 Remote Equipment	18
204 Local Commissioning	19
206 Install Tray Trunking and Conduit	20
206 SSDS(Architectural)	21
106 Wall & Floor Finishes	22
206 LL Fire Main Conns	23
206 Elec/FA Cabling & Wiring	24
205/6 Firestopping	25
106 Ceiling Finish	26
206 Light Fittings/Sockets & Switches	27
205/6 Final Commissioning	28
6.36 E&M ROOMS(JLE PLATFORMS)	29

Bar Library 1

113 TARMAC	106 WIMPEY	210 WBL	COMMISSIONING
204 GPT	206 D & S	205/6 D & S	SUMMARY
			202 WSL

Drawn by: Jim Waterworth   Dwg JLE/W/00/81/720   Rev:- A Revised 14/Dec/95   Notes: For Comment

Project Ref: g:\e&m-cons\powproj\alc\miles.pdb Printed 24/Jan/96 17:55 Border A3PCIP2.8

10

Selection of central door locking system for InterCity coaches

P F DOUGHTY
Interfleet Technology Limited, Derby, UK

SYNOPSIS This paper describes the process that Interfleet Technology (formerly InterCity fleet Engineering) undertook to develop a system of secondary door locking to be installed onto all InterCity passenger carrying vehicles. Two different system designs were produced and evaluated against criteria of cost per life saved and through life costing of the systems. Field trials proved the principle of the system for installation on to 1845 vehicles of 42 different types. This process, from the initial design studies to completion of installation, had to be accomplished within a 3.5 year period with the minimum of disruption to the passenger service.

1. INTRODUCTION

1.1 Door Open Incidents

During 1991 instances of passengers falling from trains associated with open door incidents were highlighted and as a result two actions were initiated:

InterCity commissioned a feasibility study into the fitting of an additional locking device to each passenger door.
The Health and Safety Executive (H&SE) conducted an investigation into 'falls from passenger doors' in early 1992.

1.2 Health and Safety Executive report

The H&SE published an interim report which came up with the following principal findings.

A door that was correctly closed and had the primary lock engaged could not come open in service.

During their investigation they found a number of locks jamming in the open position, although there was no actual evidence of jamming locks causing any of the door open incidents. The jamming was caused by the handle of the lock on the outside of the door being missaligned with the lock body mounted on the inside of the door. This would be a potential hazard when associated with doors jamming in the frame opening, otherwise the door would swing open and should be observed.

A system of additional locking fitted to each door would be desirable but they recognised this would be expensive.

This interim report was published concurrent with the trials of a secondary locking system instigated by InterCity and fitted to three vehicles. This system was derived as a result of the feasibility study and utilised a pneumatic bolt, operated automatically and independent of the passenger, to provide an additional means of locking the door.

As a result of the H&SE interim report and the initial trials with the secondary locking system the British Railways Board decided to fit all InterCity Vehicles with a system of secondary door locking by the middle of 1995. This completion date became a key project driver and was one of the principal criterions used in defining the system from all the options considered.

1.3   Primary Locks

Concurrent with the secondary locking project, InterCity initiated a programme for the redesign of the primary door locks fitted to all its Mark 3 and Mark 2 vehicles with the objective of eliminating the possibility of missaligning the handle and body during installation.

The design adopted for the Mark 3 vehicle doors was to mount the handle directly onto the backplate making the lock an integral assembly. With this design, and the locks being assembled in a factory, the possibility of misalignment between handle and body was eliminated.

Mark 2 vehicle doors are significantly thicker in the area of the lock than the Mark 3 door and so a similar 'Integral lock' solution was not a viable option. For this door type the drive between the handle and lock body was redesigned to cater for a significant degree of misalignment during assembly of the lock onto the door. The new drive member being shaped like a 'dogs bone' has given rise to these locks being referred to as 'dogs bone locks'.

In addition to these new lock arrangements the following initiatives were undertaken:

- Increasing the clearance between door and frame to prevent doors jamming
- Fixing fluorescent stripes onto door edges to assist station staff in detecting open doors
- Revising maintenance instructions to increase scope and frequency of inspections.

2. PROTOTYPE SECONDARY LOCKING SYSTEMS

The objective was to provide an additional means of locking the doors, independent of the primary lock and any passenger actions.

2.1 Speed Sensing System

The initial design produced in February 1992 involved pneumatic bolts being inset into each door frame, horizontally above the primary lock, and engaging with a striker plate fitted to the door edge. The system design was such that the 'fail safe' condition was with the bolts retracted and this was achieved by having each bolt closed by pneumatic pressure and opened by a return spring thus loss of air pressure would unlock the secondary locking. Air was supplied from a direct tapping in the main reservoir pipe, at each vestibule end, and through a pressure reducing valve to reduce the pressure from 10 bar to 7 bar.

The bolts were activated by a signal generated from a speed card in the Wheel Slide Protection system electronics rack. From this signal a 110v DC supply was derived and sent separately to each vestibule of the vehicle to operate the door bolt control electro pneumatic valves. The signal initiated a close door bolt function when the vehicle speed rose to 9 mile/h and opened the door bolt when the vehicle speed fell below 5 mile/h.

Apart from the derivation of the speed signal each vestibule operated independently. Each vehicle in the train formation generated its own speed signal and operated independently. No indication was given to train Crew or passengers as to the status of the system or if any of the door bolts had activated.

This system was installed in one vestibule on Mark 3 Sleeper No. 10577 for extensive testing both of the system operation involving 100 accelerations and deceleration's and for endurance over 10 000 cycles of operation. All testing was conducted at the Railway Technical Centre, Derby between March to June 1992.

As originally designed the bolt only engaged with the striker plate when the door was fully closed and if the bolt operated with the door open the door could not be closed. This was identified as a potential hazard during a risk assessment study and the striker plates were modified allow for engagement of the bolt when the door was in the safety catch position of the primary lock and also to allow the door to be closed if the bolt was extended. In order to allow the bolt to retract easily to allow the door to be closed with the bolt extended the air pressure to the bolt was reduced from 7 bar to 2.1 bar.

The risk assessment study also revealed a potential problem with respect to egress from the vehicles if the door bolts failed to retract during an emergency. A design was produced that had emergency egress handles fitted above each door which, when operated, shut off the air supply to the bolt and vented the bolt allowing the spring to retract the bolt.

2.2 Designs for Other Vehicles

The trial secondary door locking system worked off a signal from the BR Mark 2 WSP but this WSP system is only fitted to just over half the Mark 3 and HST fleets which amount to 38% of InterCity vehicles. The remaining Mark3 and HST vehicles, 32% of InterCity vehicles, are fitted with a Girling WSP system, a self contained axle mounted system which does not generate a usable speed signal. The Mark 2 vehicles do not have any form of WSP equipment at all.

In order for a comprehensive evaluation of a speed sensing type of secondary door locking system to be undertaken, designs for systems that could be installed on these vehicles were required and so 3 design studies were commissioned from Railway supply companies to:

- Derive a suitable speed signal from the Girling system on Mark 3 and HST vehicles.
- Install a suitable axelend probe and produce a speed signal for all Mark 2 vehicles.
- Investigate any and all available technologies to provide a suitable means for producing a signal for any type of vehicle.

In all cases the purpose was to produce a vehicle speed signal that would provide a supply to operate the EP valves and initiate a close bolt function at 9 mile/h rising speed and an unlock bolt function at 5 mile/h falling speed. The basic criteria for the designs were that they should be capable of being installed within 24hours, the vehicles would not have to be lifted and any maintenance required should be undertaken within existing maintenance periodicity's.

In addition to offering technical solutions for the three design areas above the Suppliers were also instructed to submit details of the design time for development of the system and an estimation of the duration and man hours required for installation.

2.3   Senior Conductor System

During the testing of the WSP prototype vehicles a decision was taken to investigate, in parallel, a system that would be under the control of the train crew and so two Mk 2e vehicles, were withdrawn from store to act as test vehicles for a Senior Conductor System.

The vehicles were modified at the Engineering Development Unit in Derby but during installation it was found that one vehicle was unsuitable for service running without significant repairs and so this vehicle was used for extensive static reliability/endurance testing.

The design remit was to have one control panel in each vestibule, at diagonally opposite corners of the vehicle. As only InterCity senior conductors would operate the system the panel was required to have the same configuration for the controls as that fitted to Mark 4 coaches. The location of the control panels would be such that senior conductors could operate the panels whilst looking out of the drop light window of the door.

Through train wires comprising UIC 12 core screen cables were installed on each vehicle and jumpers were used to connect vehicles together. The train wire cable was located within the ceiling space, strapped to existing fixtures, in order to save the time of installing conduit.

Mark 2 vehicles are equipped with 24V and Mark 3 vehicles with 96V batteries. Because Mark 2 and Mark3 vehicles work in mixed formations a common control voltage was required. The operating voltage selected for the control panel and control signals was 24V d.c.

Fig 1 shows the control panel layout. The green light on the panel showed that the panel was available for use and when initialised using a high security key the orange light was illuminated to indicate the panel was powered up. To prevent two control panels being operated simultaneously a dual panel inhibit circuit was installed so that when a control panel was initialised all other panels on the train were made unavailable and their panel available lights were extinguished.

To unlock the doors the two red buttons had to be pressed simultaneously for at least two seconds. This design ensured that the open door operation was a conscious action and that doors could not be unlocked accidentally. The unlock buttons only unlocked the doors on the same side of the train as the control panel.

Pressing the blue door lock button locked all doors on both sides of the train. locking the doors enabled the greensignal button which allowed the ready to start signal to be given to the driver on push pull and HST services.

As a guide to passengers, internal indicator lights were fitted adjacent to each door which were illuminated when the doors were unlocked to reveal the legend 'door unlocked'. Orange indicator lights were also installed externally at each door position. There was no interlocking for these lights and so a change in status of the lights only indicates that a lock or unlock signal had been received by the adjacent panel.

The door bolts were located on the headrails above the doors and, operating vertically, engage with a striker plate fixed to the top frame of the doors. This position was selected in preference to the position used on the WSP system because of significantly reduced installation times, Mark 2 door frames would not easily accommodate an edge fixing striker plate and with the modification to the striker plate to accommodate locking in the safety catch position the increased protrusion from the door edge presented a hazard of head strike injuries when the door was open.

The striker plates were fitted with mechanical indicators to show the status of the door bolt, the indicator showing yellow when unlocked and black when locked. The striker plate had two engagement areas to allow the bolt to engage when the door was in the safety catch position but the mechanical indicator only operated when the door was fully closed. A chamfered edge to the striker plate lifted an extended bolt to allow the door to be closed even with the secondary lock bolt extended.

The striker plate was initially only fixed to the top rail of the door frame but the strength of this mounting was found to be very dependant upon the extent of rusting of the door frame. The design was altered to allow the striker plate mounting to additionally locate on to the frame just above the window which gave a significant increase in rigidity

The unservicable vehicle (5841) was fitted out for reliability/ endurance testing with computer controlled testing equipment that subjected the secondary locking system to a continuous test cycle. This cycle contained a series of lock and unlock operations on alternate sides of the vehicle during which three of the doors were always closed and the fourth door was opened and closed by a pneumatic ram every other cycle.

The testing was continued until one million bolt operations were completed.

3.   SYSTEM EVALUATION

3.1   Service Running

The finalised WSP design was installed to three HST day coaches, 42324, 41064 and 42124 by the Engineering Development Unit in Derby for operation on the Midland Main Line. Only the last vehicle to be modified, 42124, was fitted with emergency egress at each door.

For the initial period in service, from 22 July until 31 July 1992 inclusive, observers were present on the train at all times. For the first three days one observer was present on each coach but this was then reduced to one observer on the train.

After this initial period the observers were withdrawn and technical riding inspectors from the Midland Main Line monitored the vehicles in traffic and reported on any defects.

The servicable vehicle (5846) fitted with the Senior Conductor System was used for service evaluation on the Midland Main Line between Derby and London. Because this was the only vehicle in the train set fitted with secondary locking and the vehicle had to be always marshalled next to the regular Senior Conductor vehicle, the period of service testing was limited.

Observers were also present on this vehicle during the trial period and their information, together with feedback from the senior conductors was used in the assessment.

3.2   Video Analysis

3.2.1 Equipment used

In order to assess the effects of the various systems on passenger behaviour a series of monitoring exercises utilising video cameras was undertaken concurrent with the trial running. This was not only done on the prototype sets but additionally on a standard unfitted set for comparison. All video monitoring was performed on the Midland Main Line.

Observations of the passenger behaviour were made using miniature video cameras mounted above four of the doors on each of the different configurations listed above. In order that passenger behaviour was not modified by them realising that they were being filmed the cameras were hidden behind silvered screens. The four Video Cassette Recorders were housed in plywood boxes installed in the base of the luggage stack at each end of the vehicles.

The cameras were activated by BR staff travelling on the trains using a radio pager to trigger all the VCR's simultaneously. The VCR's were programmed to record for a five minute period and then automatically stop. This period of five minutes was chosen to cover the train entering the station and usually the departure as well. The staff onboard maintained a log of the events triggered for each station stop.

Additional video recordings were taken, from the platform at St Pancras station, of train sets arriving in order to monitor from outside the vehicles the passenger interface with doors not fitted with central locking.

3.2.2 A summary of the findings of these studies:

There were no instances recorded of passengers interfering with the doors other than when the train was approaching a station. In view of the small number of open door incidents a year and the relatively short duration of the study this was not considered significant.

From the on train observations of the approach to stations the actions of passengers attempting to exit early i.e. before the train had stopped are shown in the following table. It was noted that the presence of a queue behind a passenger significantly increased the likelihood that the passenger would attempt to exit early.

	Standard Vehicles	WSP System Fitted	Senior Conductor System
Exit events observed	106	173	106
Open door early, attempt	64	105	47
Percentage of events	60	60	44*
Open door early, succeed	64	63	0
Percentage of events	60	36	0

* Because the Senior Conductor was always positioned at one of the doors to activate the system his presence modified passenger behaviour. Noting events where the Senior Conductor was not present the percentage of attempted early exits increases to approximately 60%

It was noted on the WSP system, where there was no indication to the passenger as to the status of the locking system, that passengers rested against the door whilst holding the primary lock open waiting for the secondary lock to disengage. On the Senior Conductor system passengers who held the primary lock open generally looked at the indicator light inside the vehicle and therefore were not resting against the door and less likely to tumble out when the secondary lock opened.

No instances were observed of passengers attempting to exit from the wrong side of the train, but again as the number of this type of incident is low this would not be expected to have been observed during this exercise.

3.3   Hazard Operability Study (Hazop)

3.3.1 HAZOP findings

A series of Hazops were conducted on the two systems, which also included assessment of possible enhancements like interlocking the bolt operation to traction or brakes.

The following residual risks were identified with the WSP system:

> Because the bolts did not engage until 9 m/h there was the potential for accidents due to late boarding passengers being able to open the doors.
>
> Passengers would still be able to open doors before the train had stopped, as shown in the video exercise.
>
> Whenever the train stopped the door bolts were released, even if the train was not at a station i.e. stopped at a signal. This would allow customers who were disorientated, or through deliberate action, to leave the train.
>
> The doors on the non platform side of the train were unprotected whenever the train is stopped at a station.
>
> Hazards associated with the Senior Conductor unlocking early, unlocking wrong platform side and failure to unlock were considered together with the liklehood of them occuring were considered.
>
> Aportioning ISO lives to the above risks, together with the poor reliability of the WSP electronic equipment, meant that the WSP system would have significantly higher estimated residual ISO lives at risk than with the Train Crew system.

3.3.2 System Enhancements

Interlocking the operation of the door bolt to either the brakes or traction power was evaluated and not considered to offer any measurable safety benefit. As either system provided only a secondary lock and the principal door retention was still the primary lock there would still be a requirement for station staff to ensure doors were closed before departure.   The hazop staff considered that the benefits to be gained by interlocking ensuring doors were closed would be offset because staff could think that the doors must be closed when the indicator lights went out and therefore not check. If this happened they may miss open doors where there was a mechanical failure with the interlocking. It was also noted that power door vehicles which have interlocking suffer significantly high failure rates with the door systems.   Analysis of the maintenance records for these failures showed that in approximately 75% of all instances there was a fault with the interlocking and not the door locking.

The use of indicator lights both internally and externally was considered to have a significant beneficial effect in providing information to passengers for both boarding and alighting. The external indicators would also be beneficial to station staff and train crew as an indicator that the door lock signal had been initiated and received by every vehicle.

Loss of battery supply on a vehicle due to MA set failure was not uncommon and so to prevent doors from becoming unprotected one of the train wires was used to power panels. This enables the central lock bolts and control panels to remain operational on up to three vehicles in the rake with flat batteries. However, in this condition the internal and external indicator lights will not be operational as they require more current than the train wire can supply.

Because interlocking was not seen to be beneficial the mechanical indication at each secondary lock striker plate to show that the bolt had activated was considered to be essential. It would be the duty of the train preparer to check these indicators to ensure that the doors were actually locking.

3.3.3 System Selection

Costings were produced for the two types of system and their various options which showed the WSP system to be approximately 30% cheaper to install.

However, because the WSP system would be a stand alone separate system for each vehicle, routine tests could only be conducted on individual vehicles. Routine test on existing WSP equipment were undertaken at the first examination where the vehicle was stopped for at least one day, approximately every two months. In order to prove the integrity of the secondary locking system it would have to be tested every night. This would significantly increase downtime, in the worst case for the Girling system this work would take 46 mins per vehicle. These tests could not have been accommodated within existing vehicle diagramming causing a significant increase in maintenance costs and a reduction in vehicle availability.

As a result of the above evaluations the Senior Conductor system was selected to be fitted, Fig 2 shows a typical vehicle equipment layout.

4. FINAL SYSTEM

Once the type of system had been selected there remained the question of how to progress the installation. There were a number of options available:

i) InterCity could undertake the detail design, procure the material and control the installation at selected sites.

ii) Two contract could be placed, one to cover design of system and supply of material to form kits of parts and the second contract to install these kits.

iii) This option was to award one contract to cover design and installation.

The resources available within InterCity were not sufficient to allow option i) to be progressed within the desired timescale.

As a result of advertising the intent to install secondary locking in the European Journal 25 firms expressed an interest and were sent a prequalification invitation to tender inviting offers for:

- System design and supply of kits of parts
- Installation of kits
- Design and installation as a turnkey project.

The firms were asked to provide indicative prices, confirmation that the required timescales could be achieved with a timeplan detailing milestones and an indication of the sub-contractors and where applicable installation sites that would be used.

17 prequalification offers were received, six made bids for design and supply of kits, two were only interested in installation and the remaining nine responded with offers to undertake design and installation.

From the analysis of the responses it became evident that there would only be a minimal financial advantage in adopting the design and installation as separate contracts. Awarding the contract as a turnkey project also reduced the risk associated with possible conflict between supplier of the kits and the installer and so it was decided to proceed with tendering on a turnkey basis.

Four firms were invited to tender and the contract was awarded to ABB Transportation Ltd in June 1993 with the following major milestones:

- First vehicles to be on works for one month and finished by December 1993
- Installation on a production basis to commence in January 1994
- Vehicles to be shopped as rakes and be completed within one week
- Up to four installation sites to be operational at any one time
- Final vehicles to be completed by June 1995

The last rake was finished and left ABB Crewe works on 17 July 1995 just 17 days behind plan despite:

- Change of vehicle ownership to the three Rolling Stock Leasing Companies
- Change in operation of the vehicles to eight Train Operating companies
- Change in operation of the Works Train network to RES
- A series of one day strikes by drivers during the summer of 1994 effecting moves

These factors caused the Interfleet project team major chalenges to the logistics of ensuring that the correct train sets were input to the correct instlation site at the correct time and the finished sets were returned to their depot.

5   IN SERVICE EXPERIENCE

The service experience to date indicates that operating staff like the system because it allows them to control the door operations and improves the entry and exit from stations. The system has acheived its reliability targets and there has not been a passenger fall from an open door on any InterCity train for 18 months.

6   CONCLUSIONS

The secondary door locking project has acheived its objectives namely;

- To be installed by the midddle of 1995
- To reduce loss of life and serious injury

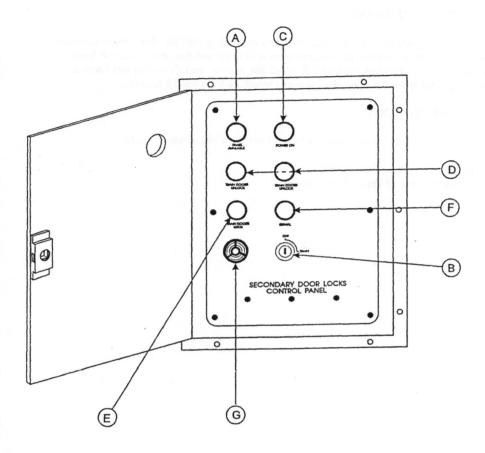

Fig 1 Secondary Door Locking Control Panel

Key: A   PANEL AVAILABLE Ligth (Green)
       B   Keyswitch
       C   POWER ON Light (Yellow)
       D   TRAIN DOORS UNLOCK Pushbuttons (Red)
       E   TRAIN DOORS LOCK Pushbutton (Blue)
       F   SIGNAL Pushbutton (Green)
       G   Buzzer

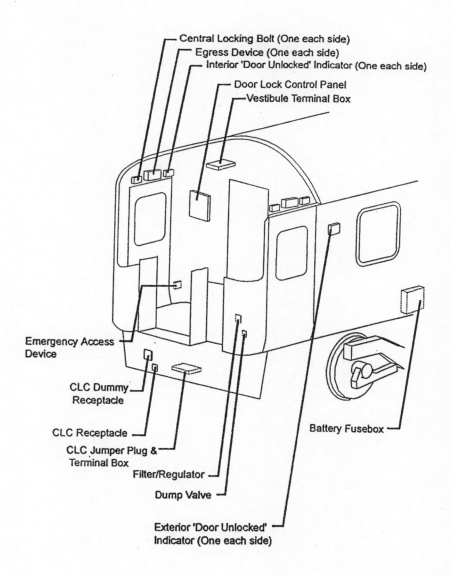

Central Locking Bolt (One each side)
Egress Device (One each side)
Interior 'Door Unlocked' Indicator (One each side)
Door Lock Control Panel
Vestibule Terminal Box

Emergency Access
Device

CLC Dummy
Receptacle

CLC Receptacle

CLC Jumper Plug &
Terminal Box

Filter/Regulator

Dump Valve

Exterior 'Door Unlocked'
Indicator (One each side)

Battery Fusebox

FIG 2: TYPICAL VEHICLE EQUIPMENT LAYOUT

FIG 2.   TYPICAL VEHICLE EQUIPMENT LAYOUT

C511/9/036/96

Improving passenger safety at platforms

T A W GEYER BSc, MErgS, SocRiskAnal, **C P CHAPMAN** BEng, MIEI, and **M I MORRIS** BA, MA, PhD
Four Elements Limited, London, UK
P R CHISLETT BA, MBA
London Underground Limited, London, UK

SYNOPSIS

This paper describes a risk and cost benefit analysis of a range of measures with the potential to improve passenger safety at the platform/train interface. London Underground Limited recognises that risk assessment can provide a systematic and rational approach to safety management, and can be used to ensure that the risks are managed and controlled to levels which are as low as reasonably practicable (the ALARP principle).

1 BACKGROUND

A number of passenger accidents at the platform/train interface, some resulting in fatal injuries, has led London Underground Limited (LUL) to consider what additional measures might be adopted to improve passenger safety in this area still further. In fact the London underground system is extremely safe at the platform/train interface - over two thousand million passenger boardings and alightings take place each year without incident. Nonetheless, whenever a passenger accident does occur, there is justifiable public concern as to whether LUL has taken, and is continuing to take, all reasonable steps to avoid passengers being exposed to undue risk; and LUL itself recognises the need for continuous monitoring and improvement where practicable.

For these reasons, LUL, in conjunction with Her Majesty's Railway Inspectorate (HMRI), commissioned Four Elements Limited to consider various measures intended to improve passenger safety at platforms, to evaluate their benefit using the technique of quantitative risk assessment (QRA), and on this basis to recommend for further consideration those measures which seem cost effective.

This report describes the approach and findings of the study, which were previously presented to the London Regional Passenger's Committee (LRPC) by LUL on 26 April 1995.

Four Elements Limited is an international risk management consultancy, which specialises in the assessment and improvement of safety in the transport and major hazard industries. For example, Four Elements was commissioned by Eurotunnel to perform a detailed and extensive quantitative risk assessment for the Channel Tunnel system in support of Eurotunnel's application for an operating certificate. Four Elements has also previously advised LUL, amongst other things on policy regarding the use of risk assessment methods, including the setting of risk criteria to establish a consistent basis for appraising safety measures.

2 HOW SAFE IS SAFE ENOUGH?

The regulation of safety in the United Kingdom is based upon the fundamental principle that risks must be reduced to a level which is "As Low As Reasonably Practicable" (the so called ALARP principle). The significance of "reasonably practicable" is well illustrated by the following English case law definition:

> *"Reasonably practicable" is a narrower term than "physically possible" and seems to me to imply that a computation must be made by the owner in which the quantum of risk is placed on one scale and the sacrifice involved in the measures necessary for averting the risk (whether in money, time or trouble) is placed in the other, and that, if it be shown that there is a gross disproportion between them - the risk being insignificant in relation to the sacrifice - the defendants discharge the onus on them."* (Judge Asquith, Edwards v. National Coal Board, All England Law Reports Vol.1, p.747 (1949)).

Thus the ALARP principle allows cost to be taken into account in determining how far to go in the pursuit of safety, so that if a risk reduction measure involves "grossly disproportionate" cost, it is not "reasonably practicable".

This principle was adopted in the Health and Safety at Work Act (1974), and is the basis of the approach adopted by the Health and Safety Executive in their regulation of the major hazard industries, such as the offshore oil and gas industry. It has also been adopted by HMRI in its application of the Railways (Safety Case) Regulations (1994).

It is against this background that LUL, as a major railway operator, aims to provide a safe, efficient and quality service to its many millions of customers. As such it requires effective decision making support to assist in the management of safety and the prioritisation of capital expenditure, ultimately to discharge its legal obligation to ensure that risks are managed and controlled to levels which are as low as reasonably practicable.

LUL recognises that the modern techniques of risk assessment can help managers make better decisions on safety issues by providing a more systematic basis for appraising safety options. This study is therefore a key part of the decision making process within LUL, whose aim is to achieve an optimum level of safety for passengers at platforms, by identifying and prioritising opportunities for investment in measures which can effectively improve safety, whilst avoiding expenditure on measures whose cost is grossly disproportionate to their benefit.

3   ASSESSING THE EFFECTIVENESS OF MEASURES TO IMPROVE SAFETY

In order to demonstrate reasonable practicability, or gross disproportion, it is becoming increasingly the norm in the railway industry as well as in the major hazard industries for the extent of the benefit of a safety measure to be evaluated using the technique of quantitative risk assessment. In essence this involves firstly identifying a comprehensive set of possible accident scenarios, and then for each scenario deriving quantitative estimates of both how often it may occur (eg, once in ten years) and how severe the consequences may be (eg, one person suffers fatal injuries). The effectiveness of the safety measure is then assessed by considering the extent to which it can be expected to reduce either the frequency or the severity of each accident scenario. Finally, the results of this calculation are incorporated into a cost benefit analysis to determine the balance of practicability.

The application of the method to this study is shown schematically in Figure 1 overleaf. In this case the relevant accident scenarios and the current level of risk at the platform/train interface have been established from the record of passenger accidents for the last few years. The accident categories considered in the study are:

- person under train
- person on track
- caught in doors
- fall between train and platform (or between cars).

The next stage involved assessment of each safety measure, to determine its effectiveness in reducing the risk. Consideration has to be given to each category of accident separately, as some safety measures are effective only against certain types of accident. In addition, both the frequency and the severity components of the risk must be addressed. This is important because the various measures address these components differently.

For example, a measure such as the use of closed circuit television (CCTV) is of benefit because, once the driver becomes aware that a passenger has fallen between the train and the platform, he will not start the train, and so the injuries suffered by the passenger will be much less severe. However, the measure is not expected to influence the frequency of incidents - passengers are still just as likely to fall between the train and the platform. Alternatively, the benefit of a measure such as platform edge doors is that it would reduce the frequency of accidents, rather than the severity of their consequences.

The risk reduction assessment process was essentially a judgemental exercise, based on the experience and expertise of LUL operations personnel familiar with day to day operations, and facilitated by a risk assessment consultant with considerable experience of the issues involved. In particular, a meeting was held where estimates of the risk reduction achievable by each measure were agreed.

In some cases human factors specialists with experience in railway procedures provided additional input where the improvement envisaged relies on assumptions about staff and passenger actions.

As regards the cost estimates, LUL provided general background data on the number of stations, trains, platforms, etc, and prepared a preliminary specification of each of the potential risk reduction measures.

In general, when assessing the net cost to LUL, it is necessary to take into account the capital cost (eg, for hardware and its installation), any increase or reduction in on-going operating costs, for example, in maintenance costs and/or staff costs, and any operational benefits derived from implementation of the measure (revenue from increased usage resulting from passenger benefits). In terms of passenger benefits, there is the societal benefit associated with the reduction in accidents, the reduction in the costs associated with delays and interruptions (as a result of reducing the frequency of accidents), but also any potential disbenefit associated with disruption caused by passenger mis-use of the safety measures (eg, mis-use of emergency stop plungers).

Figure 1: The Decision Making Process

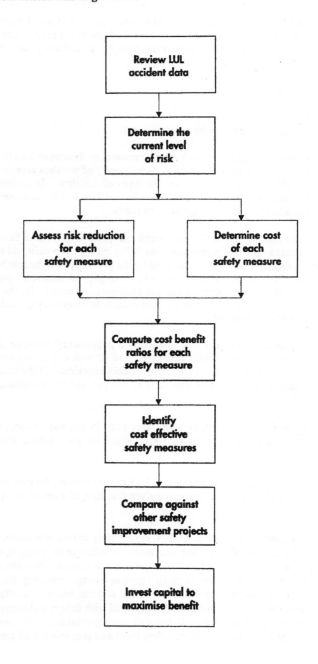

A standard cost benefit calculation was then performed for each option. LUL's preferred measure for distinguishing the merits of each of the candidate risk reduction measures is the benefit to cost ratio, calculated as the sum of passenger benefits divided by the net cost to LUL.

Although there is always a margin of uncertainty associated with the results of a quantitative analysis, experience has shown that the envelope of uncertainty is often swamped by the much larger variation in the effectiveness of safety measures themselves. Thus the benefit to cost ratio is generally a useful discriminator between upgrade options because some options are usually found to be clearly worthwhile, and others clearly ineffective, even allowing for uncertainty in the risk and cost estimates. In fact in this study, for two of the measures the overall benefit was found to be negative.

4   THE SAFETY MEASURES CONSIDERED

Each of the measures evaluated in this assessment is described briefly below.

4.1   Guards on trains
A guard is positioned at the rear of each train and has access to an emergency brake device with which to stop the train in an emergency. His primary role is to observe the doors closing and the departure of the first passenger carriages from the platform. Trains are also fitted with standard passenger emergency alarm equipment.

4.2   Platform attendants and emergency stop plungers
A platform attendant with access to an emergency stop plunger device is positioned in a prominent position on each platform with maximum available visibility of the length of the train and platform. Operation of an emergency stop plunger will illuminate red signal lights at the trackside ahead of train to warn the driver to stop immediately whilst entering or leaving a platform. On Automatic Train Operation (ATO) lines, the Victoria Line for example, the device will interrupt track circuit signalling to bring the train to a halt automatically.

4.3   Emergency stop plungers on platforms
A series of platform wall mounted alarm devices are positioned at predetermined intervals along the platform for use by passengers and staff.

4.4   Emergency stop plungers on trains
An alternative to installing emergency stop plungers on the platform wall would be to install them on the outside of the train. This would be a simpler device and quicker acting as it could directly apply the train brakes, and would not have to illuminate red signals or interrupt ATO codes.

4.5   Train borne passenger emergency alarm automatic braking system
Fitting all trains with a (Central/Jubilee line type) modified passenger emergency alarm system which, if operated whilst leaving the platform area, will automatically apply the emergency brakes and bring the train to a halt. Beyond the platform area, the system will function as normal, alerting the driver to stop at the next station.

4.6 Reducing gaps between platform edge and doors

Platform gaps modified to comply with a maximum gap criteria (150 mm both vertical and horizontal).

4.7 Inter car barriers

A simple, robust and flexible barrier (either a gate/screen type or a simple rope) is installed between all cars to prevent passengers from falling between cars or, in the case of the visually impaired, mistaking the gap for a door aperture.

4.8 CCTV monitors in cabs

A Central Line type CCTV facility is provided in all operational cabs to enable the driver to observe the approach and departure of a train from the platform.

4.9 Platform edge doors

Doors are provided which run the full length of the platform and cannot be readily climbed by passengers. The doors can only open if a train has properly berthed and must be proved to be closed before a train departs. A door may need to be opened manually from the train side if normal power falls.

4.10 Platform surveillance using CCTV

Sufficient station staff available to monitor all departing trains from a central point in the station. This measure would necessitate enhanced communication systems to ensure that the station staff monitoring the CCTV have a means of immediate contact with the driver in the event of an incident, or some means of automatically stopping a train, for example, access to an emergency stop plunger.

4.11 Improved door engineering/control mechanisms

Reducing the doors closed tolerance to the latest standard - 6 mm. All rolling stock would be fitted with door closing alarms, and all door system tolerances improved. Door edge rubbers would be stiffened to the maximum practical extent.

5 FINDINGS AND RECOMMENDATIONS

Each of the measures described above has been assessed, and the findings and recommendations are described below. A scatterplot summarising the results of the cost benefit analysis is shown in Figure 2.

1. Platform emergency stop plungers for use by passengers, and emergency stop plungers fitted on the outside of trains, overall result in a negative benefit to passengers.

2. Reduction of gaps between platform edges and trains, and improvements to train door engineering and control mechanisms generate only very small improvements in safety (because they each address essentially only one category of accident), and are not cost effective.

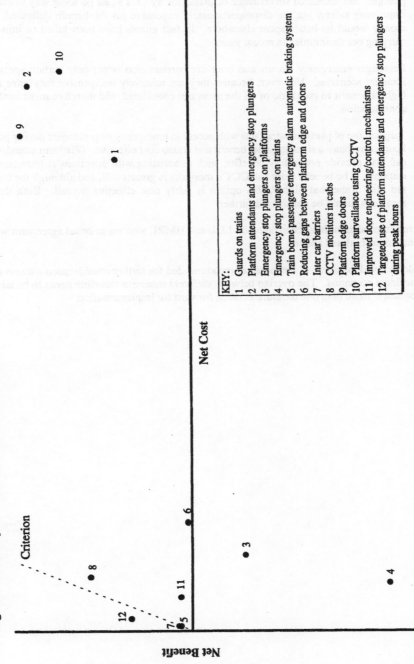

Figure 2 : Scatterplot - Net Cost Versus Net Benefit

Net Benefit

Net Cost

Criterion

KEY:
1 Guards on trains
2 Platform attendants and emergency stop plungers
3 Emergency stop plungers on platforms
4 Emergency stop plungers on trains
5 Train borne passenger emergency alarm automatic braking system
6 Reducing gaps between platform edge and doors
7 Inter car barriers
8 CCTV monitors in cabs
9 Platform edge doors
10 Platform surveillance using CCTV
11 Improved door engineering/control mechanisms
12 Targeted use of platform attendants and emergency stop plungers during peak hours

3. Platform edge doors, guards, platform attendants with access to an emergency stop plunger, and increased surveillance of platforms by CCTV, all go some way towards improving safety, but are disproportionately expensive for the benefit delivered, so money would be better spent elsewhere. In fact guards have been killed or injured carrying out their duties in recent years.

4. Passenger emergency alarms and inter car barriers also target only certain specific accident scenarios. However, because they are relatively inexpensive they have the highest benefit to cost ratio of all the measures considered, and therefore merit further consideration.

5. Targeted use of platform attendants with access to emergency stop plungers during peak hours contributes a significant net benefit and is also cost effective. (Platform attendants can also provide non-safety benefits, such as assisting with directions at interchange stations). The benefit from cab CCTV monitors is greater still, and although the costs are also somewhat higher, this option is fairly cost effective overall. Both these measures should be considered further.

These recommendations have been noted by LUL and HMRI, who are in broad agreement with the conclusions outlined above.

It should be noted that some of the measures recommended for further consideration address the same accident scenarios. The overlap between different measures therefore needs to be taken into account if more than one measure is taken forward for implementation.

C511/9/051/96

Risk assessment for Class 373 Eurostar trains on 750V dc and 25KV ac infrastructure in the UK

V NENADOVIC BSc, MSc, CEng, MIEE
British Rail Projects, Risk Assessment Group, London, UK

Synopsis

Class 373 3-phase drive trains, although very similar to the established French TGV high-speed trains, present some new and unique challenges for safe high-speed operation over the railway infrastructure in the UK. The designated Class 373 routes were electrified and re-signalled at different times since 1965 using evolving design standards. It is now mandatory to present a Safety Case for running on UK routes. The ongoing risk assessment is identifying, quantifying and resolving the train compatibility hazards on these routes. · The electromagnetic compatibility (EMC) with the existing signalling equipment is the key safety issue. This paper describes the strategies adopted for EMC risk assessments, application of safety targets, and recommendations for immunisation of the infrastructure.

1   Introduction

1.1 Background

The railways of Britain, France and Belgium signed a contract in December 1989 for the initial fleet of 30 high speed trains to operate international service through the Channel Tunnel.   The main objective for the operators of the Class 373 trains, each able to accommodate 760 passengers, is to provide a level of safety, speed, comfort and reliability that can challenge the airlines on the Europe's busiest routes.

The Class 373 is one of those rare trains that have been designed with capability to operate on railway systems in the UK (750V d.c. third rail, and 25kV a.c.), France (25kV a.c. two modes), and Belgium (1500V d.c. over-head catenary). Based on the proven 300km/h French TGV train, with its powerful 3-phase traction package and sophisticated computer control systems, the Class 373 has to be fully compatible with the significantly different power supplies, catenary arrangements, signalling and telecommunication systems in the three countries.   Having successfully resolved most civil, mechanical and electrical interface problems, the UK operators were faced with the complex electromagnetic interference (EMI) risk assessment.

The compatibility problems of 3-phase drives and the railway infrastructure have been recognised since the early 1980's, when systems like the Maglev between Birmingham Airport and Birmingham International Station/NEC entered service (Refs. 1,2). The potential hazard of earth faults on the Maglev's 600V d.c. supply rails affecting BR track circuits was recognised and resolved by providing reliable substation protection.

To understand and define the key parameters that affect the safe operation of Class 373 in the UK, it was necessary to undertake in 1994 a series of tests, conduct electromagnetic interference studies, and assess the risks to signalling and telecommunications systems. A risk assessment process was developed, using formal methods proven in the petrochemical, offshore and nuclear industries, that satisfy the Railway Safety Case Regulations (RSCR) and safety approval practices in Railtrack (Refs. 3,4).

Demonstration of compliance documents produced by the train manufacturers and operators were based on the railway industry EMC standard BR 1914 (which is no longer mandatory since the introduction of RSCR) and advice of experts. This approach, together with established infrastructure design, installation and maintenance procedures, required thorough re-examination to identify and quantify areas of risk to safe operation of the new trains.

1.2 Class 373 Project

The original work in the risk assessment of trains employing 3-phase traction packages was started by the train manufacturers, BR Research, Signalling Control UK, and experts from other BR organisations (Ref. 5). In June 1994 it became apparent that an integrated project team is necessary to pull together the signalling, traction and safety engineering expertise, and achieve the safety certification timetable required by the train operator, European Passenger Services (EPS).

The project management of this complex process by British Rail Projects, from initial feasibility studies to revenue-earning operation, was one of the key factors in achieving the fleet safety certification of Class 373/1 on existing DC infrastructure in November 1994.

Computer models of the traction package, signalling equipment, track circuits, substations and power distribution systems had to be developed to simulate many different configurations expected in normal operation and under various failure modes that would be impractical to analyse by traditional methods and tests. The very demanding safety targets for individual equipment (10^{10} hours MTBF which could cause a wrong side track circuit failure) required careful application of conventional probabilistic analysis to avoid meaningless conclusions.

The efforts of many organisations and individual experts had to be co-ordinated and presented in the form of survey results, test reports and safety assessments that can stand up to independent review. These findings were further subjected to corroborative verification by tests and approval by Raitrack EE&CS Safety Approval Panel (SAP). The project team concentrated its efforts first on the Channel Tunnel Routes from Waterloo International Station, and obtained a safety certificate for restricted operation of Class 373/1 by August 1994, and for unrestricted fleet service by November 1994. The team then embarked on the detailed EMC risk assessment of the 25kV a.c. WCML and ECML routes for Class 373/2. This ongoing risk assessment work is the main subject of this report.

1.3 Related Projects

A considerable exchange of information is taking place between a number of 3-phase traction projects in the UK despite some real and imaginary commercial fears about confidentiality and "intellectual property rights". Unfortunately, with some notable exceptions, the manufacturers are very reticent to disclose full EMC data.

The safety cases are at this time being prepared by a number of different organisations (Table 1).

Rolling Stock	Type	Routes	Manuftr.	Opertr.	Safety Case
Class 373/1 Eurostar	high-speed train 2 locos + 18 cars	750Vdc CTR's NPID, D. Moor	GEC Alsthom	EPS	British Rail Projects (completed)
Class 373/2 Eurostar	high-speed train 2 locos + 16 cars	750Vdc CTR's NPID, D. Moor 25kVac WCML 25kVac ECML	GEC Alsthom	EPS	British Rail Projects (cleared for dc routes)
Nightstock (various locos)	sleeper train 2 locos+2 gen vans + 18 cars (max.)	750Vdc CTR's 25kVac WCML non-electrified	GEC Alsthom	EPS	British Rail Projects
Class 92	freight locomotive & Nightstock	750Vdc CTR's 25kVac WCML	ABB	RfD/EPS	British Rail Projects
Class 365 Networker Express	EMU	750Vdc Kent Coast Serv.	GEC Alsthom	Eversholt Leasing	British Rail Projects
Class 365	EMU	25kVac West Anglia Serv.	GEC Alsthom	Eversholt Leasing	British Rail Research
Class 325 Royal Mail Railnet	EMU	all 750Vdc and 25kVac routes	ADTranz (ABB)	Post Office	ADTranz (ABB)
Class 323	EMU	25kV routes (dc routes later)	Holec	ROSCOs	Holec (cleared around Birmingham +Manchester)
Heathrow Express	EMU	25kV Heathrow to Paddington	Siemens	BAA	British Rail Research
Juniper	EMU	all 750Vdc and 25kVac routes	GEC Alsthom	?	?

Table 1 - Related 3-phase Train Projects

2   Train Interference

2.1 General

The Class 373/2 train, comprising 16 coaches and two permanently coupled locomotives with a combined rating of over 7MW and weight of 793 tonnes, is one of the most powerful sets in the UK. The train has 12 motored axles; each of the six motors delivering 1MW. When running in France the rating of the train is increased to its full design capacity of 14MW.

At a cost of £25 million each, Class 373/2 employs some of the most advanced designs, safety features and protection systems to ensure safe, reliable operation. In particular, the crashworthiness, door systems, braking systems, communications and drivers controls exceed the requirements of all current standards.

The Class 373/2 has d.c. auxiliary supplies 530V and 72V that are bussed along the train. If an earth fault occurs, there is a possibility that d.c. current could circulate through the train body and the axles falsely energising the track circuit.

The Class 373/2 is equipped with a high-integrity differential current monitoring device (DCMD) that detects any earth faults and disconnects the faulty supply. This is set to operate at a current level tolerable to DC track circuits.

The traction package EMI has been minimised by careful design of the traction package, with further protection provided by an automatic interference current monitor (ICMU) having a very high reliability of 5×10^9 hours WSF MTBF. Unfortunately, the ICMU cannot provide protection from the effects of transformer inrush which was found to be the main cause of interference.

2.2 Transformer Inrush

All a.c. locomotives and multiple units are equipped with a transformer to convert the 25kV supply voltage to lower voltages which are compatible with the traction motors and control gear. There is concern that the inrush current of transformers, used on the new more powerful trains, could be significantly higher than that on the existing rolling stock. This effect may be severe enough to cause false clearing of track circuits occupied by a train (Wrong Side Failure, or WSF), or false indication of occupation of unoccupied track circuits (Right Side Failure, or RSF), on or around the line of route.

3   The Railway Infrastructure

3.1 750V d.c. System

The significant sources of 50Hz interference, affecting many vane type 50Hz track circuits in the former BR Southern Region, are the mains ripple voltage produced by the traction current rectifiers in substations and 50Hz train filter ringing at rail gaps. False operation of these track circuits, also known as VT1 after the most common type, was prevented by interference current monitors (ICMU) that disconnect the traction supply when excessive level and duration of 50Hz interference is detected. However, the conductor rail gapping caused an unacceptable number of spurious trips even with the highest allowable ICMU setting. It was then necessary to desensitise the affected track circuits by modification to VT1-SP (slow to pick), or replacement to more immune type, e.g. HVI, TI21 or FS2600.

Parts of the route were equipped with reed track circuits operating at frequencies of 363 to 423Hz. Since the train was producing EMI in this range, all reed track circuits had to be removed. Other types of track circuits were not affected.

The extent of above conversions on the branch lines away from the main route depends on the propagation and attenuation of EMI produced by the train. These interference currents circulate through the rail network, earth and substation rectifiers. Detailed modelling of this network confirmed that the practical limit of interference effects is generally two substations beyond the main route. An arrangement for substation switching was introduced to mitigate during substation outages.

3.2 25kV a.c. System

Electrical supply to the train is via the 25kV overhead catenary, current being collected by a carbon composition rubbing strip on the train pantograph. Traction current is returned via the common rail, or through the ground. Some schemes have booster transformers to draw the return current into the rails. Recent schemes use booster transformers, arranged to draw the return current into a return conductor, thereby minimising the amount of EMI induced in communications and other circuits running parallel to the railway. Return path is therefore very varied, and may be through the rails, return conductors, or the ground, depending on the electrical arrangements, ballast and ground conditions. The contact wire is split into sections to maintain an insulated gap between electrical sources, such as National Grid supplies and phases. Electrically discontinuous gaps at Neutral Sections, where the power to the train is disconnected whilst the train passes from one supply to another, cause inrush.

4 DC Track Circuit

4.1 General

DC Track Circuits provide detection of trains in a section by virtue of the train axle electrically shunting the d.c. voltage, that is applied to the running rails at one end, and preventing this voltage from reaching the relay, also connected to the running rails at the other end. Only if the relay has a voltage above a threshold will it energise to make its contacts giving a means of detecting an unoccupied track circuit.

Adjacent track circuits are electrically isolated by Insulated Rail Joints (IRJs) in the signal rail and are fed in the opposite polarity to their neighbours so as to detect a failure of the IRJ. One rail is electrically continuous for the purpose of carrying the traction return current, however it is also shared by the track circuit. It is this sharing of the common rail that provides a path for the train produced interference to be coupled to the track circuit.

At least one slow-to-pick Track Proving Relay (TPR) follows the Track Relay (TR) on the designated routes. This provides a delay of about 400 ms before the track circuit can be energised with transients presented to the TR. The contacts of the last TPR relay are used to feed the signalling circuits, interlocking and indication circuits. It is the false energisation by EMI of last TPR that is considered as a wrong side failure of the track circuit.

4.2 The Failure Mechanism

The traction current passing along the common rail develops a longitudinal voltage along the rail in proportion to the impedance of the rail.. This voltage is applied to the track circuit rail by a pair of wheels short circuiting the rails between the track circuit feed unit, and the relay. The effect is greatest when the axle is immediately adjacent to the track feed unit, and is reduced if the traction return current finds an alternative path through cross-bonding, earth leakage, or a return conductor connection.

A Wrong Side Failure (WSF) can occur if this voltage contains a DC component sufficient to energise the last TPR (or indeed prevent it dropping) when the train is occupying the track circuit.

A Right Side Failure (WSF) can occur if the voltage opposes the legitimate track circuit voltage, causing the track relay to drop when no train is present. This will generally be less than that required to cause a WSF, since the voltage requires to exceed only the difference between the Pick-up and Drop-away voltages of the relay.

5   Modelling

5.1 General

In order to quantify more accurately the effect of the inrush current produced by the Class 373/2 on various DC track circuits, a model was developed using proprietary modelling software PSpice and MathCAD. The basis of the model was a generic L, C and R lumped component circuit representation of the railway infrastructure for single and double track routes. This included all the main system parameters relevant to track circuit EMC studies:
- rail impedance including variation with frequency
- resistance and capacitance between rails and to earth
- resistance of the ballast

The DC relay was modelled as a two time constant component with parameters obtained from laboratory tests. The input to the simulation was the actual transformer inrush waveform digitised to capture d.c., 50Hz, 100Hz and 150Hz frequency components.

The modelling process (Fig. 1) requires agreement of SAP Mentor at critical stages, and relies heavily on practical tests for input data and validation of results.

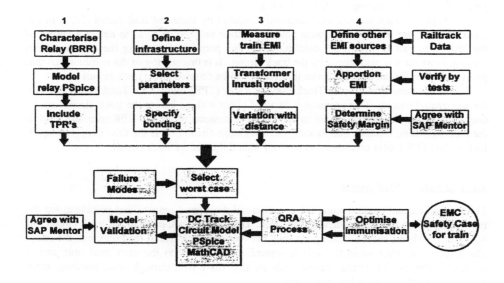

Figure 1 - The Modelling Process

5.2 Results

The modelling analysis calculated the permissible length of DC track circuits that are terminated in 9Ω QTA2 relays to BR Spec 966F2. The results are presented in graphical form (Fig. 2) to simplify evaluation of practical route situations.

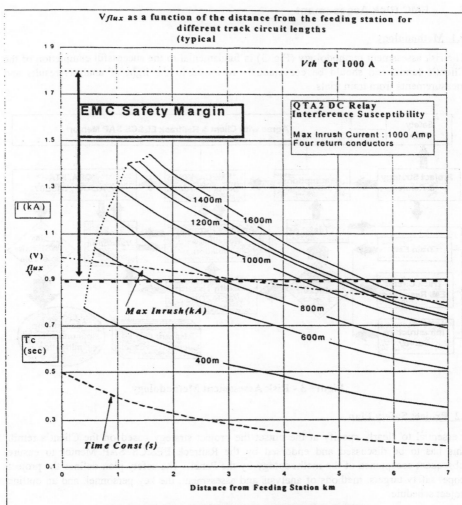

Figure 2 - Results of DC Track Circuit Modelling

The graph (Fig. 2) shows that the permissible track circuit length, provided the curve is below the *Vth* threshold including the 50% safety margin, increases as the track circuit is further away from the feeder station. This was found to be the case because the inrush decreases in amplitude as the train is further away from the feeder. This effect is shown by the dotted curve *Max. Inrush(kA)*. Another effect is that the decay of the inrush is faster if the train is further away from the feeder. This effect is also shown as a dotted curve labelled *Time Const(s)*.

6   EMC Risk Assessment

6.1 Methodology

The risk assessment methodology (Fig. 3) is fundamental to the successful completion of the Client's remit, and should be continuously reviewed in the light of analysis results and measurements from train trials.

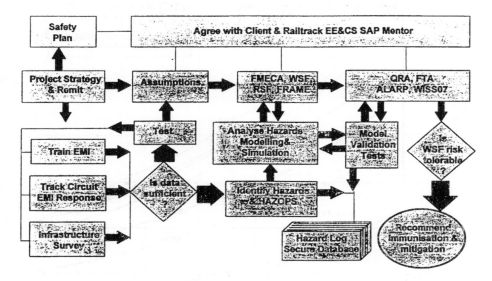

Figure 3 - Risk Assessment Methodology

6.2 Project Safety Plan

It essential to clearly specify at the outset the project strategy based on the Client's remit. This has to be discussed and endorsed by the Raitrack EE&CS SAP Mentor to ensure subsequent acceptance of the methodology by the Panel. The safety plan defines the project scope, safety targets, methods of analysis and assessment, the key personnel, and an outline project schedule.

6.3 Assumptions and Constraints

Before all survey and test data are available, it is necessary to work with a set of reasonable assumptions about the infrastructure, the train EMI and the track circuit EMI response. The assumption are agreed with the Client and Railtrack before the work starts, and then undergo regular updates as the information becomes available.

The constraints have to be explicitly stated to avoid misinterpretation of assessment results, and could include:

- limitations on train operation, e.g. speed or frequency of service
- state of the infrastructure, e.g. additional inspection of bonding
- environmental factors, e.g. no icing or night operation only

6.4 EMI Tests

The test results are an essential prerequisite for any credible risk assessment. The series of tests on the train start with the manufacturer's type tests, and continue throughout the assessment duration feeding valuable performance data for use in EMC analysis. Any tests on the track must include a co-ordinated set of measurements on the train for generated EMI, and on the ground for signals coupled into signalling systems and track circuits. The results must be carefully analysed to identify any anomalies due to other traffic or specific conditions during the tests, e.g. icing on the catenary or unusual arrangement of traction power supplies. The test reports and all test data are an important part of safety case documentation for the train.

Another important role for the on-track tests is to validate the EMC calculations and circuit models. Without this validation the modelling and simulation results are no more than interesting intellectual exercises. All safety critical technical decisions and risk assessments must be supported by actual test results. It is essential that the configuration control is strictly maintained for the train during testing because it is precisely then that changes to train hardware and control software are introduced. Any modifications have to be recorded and separately evaluated.

6.5 Hazard Identification, Analysis and Resolution

To effectively identify all hazards that could compromise the safety of a railway system, it is necessary to implement a systematic, structured approach similar to that defined in the Chemical Industries Association HAZOPS Guide. It is essential that the HAZOPS process takes place in formal, recorded meetings comprising recognised experts in the topics under discussion and an experienced Chairman. He has to act as a facilitator of the creative thinking process, ask the "what if" questions, stimulate the communication between experts, and encourage a free flow of ideas. The following steps describe the process:

1. Definition of how the train and infrastructure are intended to operate safely.
2. Characterise the main safety critical equipment to be later verified by test.
3. Identification of deviations by systematic examination of key parameters.
4. Formulate main causes as initiating events for the deviations from normal operation.
5. Define and study the consequences of deviations.
6. Identify hazards as those consequences that can cause damage, loss or injury.
7. Record the credible hazards and their resolution in a secure Hazard Log Database.

6.6 Failure Mode, Effect and Criticality Analysis (FMECA)

The FMECA is a systematic procedure for identifying the modes of failure and for evaluating their consequences. The main purpose of an FMECA is to ensure that all conceivable failure modes and their effects on operational success of the system have been considered. The basic questions to be answered by an FMECA ere as follows:

1. How can each equipment conceivably fail?
2. What mechanisms might produce these failure modes?
3. What could be the effects if the failures did occur?
4. Is the failure in the safe or unsafe direction?
5. How is the failure detected?
6. What inherent provisions compensate for the failure?

To be of any practical benefit, FMECA has to cover the whole railway system, and not just the train or the track circuit. It is important that the full consequences of a failure are established by tracing its effects methodically throughout the system. A great deal of valuable information is lost if the system is considered to comprise a number of isolated units.

The system and its mission must be defined, its operation described, failure categories identified and environmental conditions specified. Interfaces should be clearly defined. The severity of each effect on the overall performance, when coupled with the failure rate, identifies its criticality. The railway system hierarchy (Fig. 4) represents only one non-exhaustive layer, describing physical equipment, of a multidimensional picture. The other layers could describe the design, manufacture, operation, maintenance and software associated with each equipment. Failures and errors are possible at each level of this complex structure. There could be random and systematic faults as well as human error. Some of them will be common-mode failures, others will be dormant faults revealed under certain set conditions. Of particular concern are cascade failures that can lead to chain type events affecting several different areas or systems. All of them have to be identified, their consequences analysed, and their potential to cause an accident evaluated. This is essentially the work of the risk assessment team.

An inordinate amount of time and effort could be spent in applying FMECA to a real railway system. This is not the intent. Good engineering judgement must be used to determine when and how to terminate the analysis. The following indications are useful:
1. If the consequences of failure are not severe in comparison with others identified.
2. If the causes of failure are clearly incredible.

Figure 4 - Railway System Hierarchy

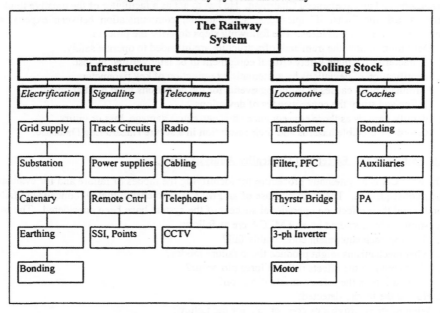

© IMechE 1996 C511/9/051

It is important to distinguish the risk associated with a WSF of a track circuit and the risk of a railway accident. A whole series of coincident, random, unlikely events has to be postulated to create a realistic accident scenario. For example, a typical FMECA event list might contain the following:

- Train traction system has a serious unprotected failure and produces excessive EMI
- Train is at a vulnerable position on the track, e.g. reduced track bonding, failed IRJ
- Power system conditions unfavourable, e.g. failed booster, emergency supply feed
- The affected track circuit used for route setting that could lead to an accident
- Another train present and exposed to danger
- The EMI persists at critical amplitude, frequency or phase for long enough
- Track circuit inherent immunity to EMI is reduced, e.g. wrong setting
- No RSFs (which are much more likely to occur due to EMI) have been reported

7   Recommendations

7.1 Safety Management Process

The whole safety management process is still developing in the railway industry, which itself is going through a period of fundamental change with privatisation, restructuring and regrouping of British Rail companies, the consultants and manufacturers. One principle, however, has been firmly established : the party which creates the risk is responsible for managing that risk (Ref. 3).

The existing data and standards relating to EMC of signalling systems and new 3-phase trains are in most cases not adequate for the depth of analysis and risk assessment required to meet the very demanding Railtrack safety targets. It is therefore important to document the methodologies agreed with Railtrack, on Class 373 and other current projects, in preparation for the work that is necessary to establish new industry standards. The following aspects of the safety process require addressing:

- reliability of infrastructure data collection methods used in risk assessment
- availability of EMC pertinent information from FRAME database
- the role of independent review

7.2 Safety Targets and Margins

It is generally agreed that the present safety targets, in the range of one million years mean time between wrong side failures of a track circuit, are excessively onerous. Even if these were achievable with the available technology, the time required to analyse all the failure modes exhaustively, and the cost of the necessary infrastructure improvements would keep most of the new 3-phase trains in mothballs for a few more years. Some of the possible areas for improvement are:

- examine apportionment of EMC to safety targets for accidents and fatalities
- define the EMI performance of existing rolling stock and increments by new trains
- involve train and signalling manufacturers
- configuration control management

7.3 Further Analysis and Tests

The theoretical work on characterising the susceptibility of signalling infrastructure to EMI must be corroborated by practical tests to remove any unnecessary pessimism in various assumptions that had to be made. In particular the following problems require better understanding and definition:

- saturation effects in booster transformers during high inrush currents
- the proportion of train EMI actually coupling into a track circuit
- short pulse response of track circuit relays

8 Conclusions

The considerable challenge of operating fast, efficient and safe 3-phase drive trains on the 1965 electrification and signalling systems in the UK has been faced by a number of projects recently. The incompatibility of some track circuits with the electromagnetic interference produced by the new trains has become a major cause for concern. To resolve these potentially serious interface problems, the Class 373 risk assessment team initiated a series of theoretical studies, hazard analyses and tests. The results of this work enabled Railtrack to provide fleet safety certification for operation of Class 373/1 on the 750V d.c. network between Waterloo International Station and Channel Tunnel in November 1994. The work has since continued on the risk assessment of the 25kV a.c. routes from London to Manchester and Glasgow. A significant effort has been applied to the definition of transformer inrush effects on DC track circuits which predominate on these routes. Some technical issues still remain to be resolved by practical tests. It is recommended to re-evaluate the safety targets and margins with Railtrack to avoid further delays in putting the new trains into operation.

Acknowledgements

Thanks are expressed to many colleagues in British Rail Projects, British Rail Research, Railtrack, Signalling Control UK and European Passenger Services whose work is described in this paper and who contributed to the text and results presented.

References

1. Johnson W., Nenadovic V., "Word's First Maglev Operation Moves into the Test Phase", Railway Gazette International, April 1983.
2. Nenadovic V. and Riches E.E., "Maglev at Birmingham Airport", GEC Review, Vol. 1 (1), 1985.
3. Remit for Production of a Safety Case for Class 373 Signalling Infrastructure, NSE S&T Eng. (Systems and Standards), Ref. SR1-373S, v.1, 21 May 1993.
4. Overview of Safety Certification for New Traction and Rolling Stock, NSE S&T Dev. Eng., Ref. NS1-T&RS, v.2, 5 February 1993.
5. Moore I.T., "Signalling Infrastructure Safety Cases for Channel Tunnel Services over British Main Lines", IEE Conf. Electrified Railways in United Europe, Amsterdam, March 1995.

C511/9/082/96

Building your railway safety case – feedback from the front line

J P STEAD BSc, MSafRelSoc
EQE International Limited, Cheshire, UK

SYNOPSIS

This paper addresses the major issues involved in planning, constructing, submitting and maintaining a Railway Safety Case. The emphasis throughout is on practical advice which can be applied by various types of organisations to assist in the whole life cycle of railway safety cases. The paper draws upon practical experiences gained whilst working with a variety of train operators and railway infrastructure organisations over the period in which the current safety case regime has been developed.

The objective is to assist future efforts in avoiding the technical, engineering, administrative and managerial pitfalls which may be encountered. Calculation of risks, comparison with railway group safety targets and the principle of ALARP are discussed. It is hoped that the paper will help to dispel some of the myths surrounding this subject.

1. INTRODUCTION

At the beginning of April 1994, three new pieces of legislation were brought into force under the Health and Safety at Work Act 1974. These were the Railways (Safety Case) Regulations (1), the Railways (Safety Critical Work) Regulations (2), and the Carriage of Dangerous Goods by Rail Regulations (3). This first of these pieces of legislation forms the topic which this paper addresses.

Since April 1994, all organisations intending to operate a railway system, or part of such a system, in the UK, have been required to produce a safety case document detailing how safety will be achieved and maintained for the duration of the operation. In this context, safety applies not only to those working on the railway, but with equal importance it applies also to railway passengers, and to the general public who may interface with the railway at certain points, for example stations and level crossings.

Both train operators and infrastructure owner-controllers are covered by the Railways (Safety Case) Regulations, and with specific exceptions this legislation applies to all standard gauge heavy railway systems and a great many minor railways which are operated by independent companies and volunteer groups throughout the country. The principal infrastructure owner-controller in the UK is Railtrack. Railtrack's own safety case has been submitted to the Health and Safety Executive (HSE) and has been approved. In turn, each of the separate train operating units (TOUs) formed by the dissolution of the British Rail organisation has prepared and submitted its own safety case to Railtrack for consideration.

Many safety cases have now been written, submitted and approved. In the course of this activity common problem areas have become apparent. The rest of this paper is devoted to explaining how such difficulties may arise, and showing that in most cases they are simple to avoid if a safety-conscious organisation really does lie behind the paperwork.

2.   THE BASIC COMPONENTS OF A SAFETY CASE

Railtrack has deliberately left the format of each safety case to be decided by the TOU itself. This is to allow each organisation maximum freedom in presenting its safety-related principles and the structure of its intended operation. However, there are generally considered to be five essential elements of a railway safety case, as outlined below, and the overall purpose of the document is therefore to bring all these elements into a cohesive whole which fully details how safety will be achieved and maintained for the duration of the intended operation.

It is vital that the safety case is structured in such a way that it can evolve with the organisation. Such a 'living document' forms the foundation of the organisational safety philosophy, and must never be allowed to gather dust on the book shelves of unoccupied offices. It must be available to all, not just a privileged few in upper management, and will preferably form part of the 'welcome pack' for new appointees. A quality assured system for document issue and control will greatly assist the safety case distribution and updating process.

The following five components are considered to be the essential ingredients of a railway safety case:

- system boundary and technical description
- hazard identification
- risk assessment
- safety management system
- summing up statement of safety

Potential problems associated with each of these elements are discussed in the following sections of this paper.

3.   THE SYSTEM BOUNDARY AND TECHNICAL DESCRIPTION

It is vital that the safety case begins with a clear statement on the boundary and operating parameters of the railway system whose safety is about to be described. For railway infrastructure, there will be geographical limits, including both route and track lengths. This geographical area may cover stations and depots, or they may be separately covered by their own safety cases. Wherever possible, it is positive to include quality diagrams and route maps within this opening section of the safety case to illustrate the wording.

For TOUs, there will be specific classes and combinations of traction and rolling stock (T&RS), and also operating times which may exclude certain parts of the day when infrastructure maintenance or repair work is being carried out. Each class of T&RS will have speed limits and loading maxima for the infrastructure on which it is planned to operate. All of this data should be provided.

It is most important that the system boundary is clear since the rest of the document will be read in the context of this boundary. It is similarly important that all information provided for assessment reflects the very latest position. There have been safety cases submitted which initially included out of date route maps, leading to ambiguity between the system boundary described in the text and the diagrams which were supposed to be supporting it. Such a beginning does not smooth the passage of the safety case through the approval process.

If there are any particular exclusions or special cases of operation applying to the railway system covered by the safety case, then these should be stated. For example, it may be that one class of T&RS is used solely for mail or parcels traffic, and so the risks to passengers arising from this part of the railway system is nil.

Once the system boundary has been established, a detailed technical description of what remains within that boundary can be presented. To some extent this data will confirm that which has already been stated. The objective should be to present all of the technical data which may need to be called upon in subsequent sections of the document to demonstrate that safety is achievable and maintainable. Engineering diagrams, line drawings and illustrations will greatly assist the reader to understand the concepts which are being described.

4.   HAZARD IDENTIFICATION AND RISK ASSESSMENT

Safety is the science of reducing risk. In order to reduce risk it is essential to know what factors import risk into the railway system, so that they may be eliminated or mitigated. Factors which exist within, or may be brought into, the railway system to generate risks are termed hazards. It is very important to note the distinction between a hazard and a risk; a hazard simply denotes the existence of a potential problem, and cannot be quantified. A risk is the quantified result of a hazard which exists. Thus the railway safety case must be able to identify all the hazards which are present before any assessment of the risks can be made.

4.1   Hazard Identification Techniques

There are a number of formalised techniques available to facilitate hazard identification, such as Hazard and Operability (HAZOP) studies, structured brainstorming, safety audits, etc. (4). Most of these have been adapted from existing techniques already commonplace in other safety-conscious industries such as the nuclear, petrochemical and offshore oil and gas industries.

Whichever technique is selected, there are a number of areas of hazard potential which can be overlooked by the unwary. Typically, these might include gaps in communication chains under abnormal situations, for example failure to clearly identify who is responsible for identifying the external emergency services in the event of a train collision or derailment. It is often also assumed that railway personnel will react in a highly efficient and logical manner when faced with a crisis situation despite being given no specific training to cope with such a situation. In this latter case it is important to record the hazard resulting from 'less than ideal' behaviour, and then action someone to consult a human factors specialist, who will be able to provide a more realistic model of staff behaviour under conditions of extreme stress, and perhaps devise a training course to prepare them for such an event.

It is beyond the scope of this paper to cover the wide variety of available techniques and to suggest which of these might be most applicable in a particular situation. Instead, hazard identification using one of the most common formal techniques, HAZOP study, will be examined as an example within this work.

4.2   Hazard and Operability Studies

HAZOP study requires a meeting involving a team of organisation 'experts' covering aspects of the system to be addressed, under the facilitation of a HAZOP Chairman and HAZOP Secretary. The most effective HAZOP study tends to take place when the facilitators are independent of the railway organisation, and therefore unbiased and apolitical. The HAZOP Chairman controls and guides the meeting, but expects the team of experts to think creatively and laterally about the potential hazards which may be present in the system.

The facilitators will have done their homework and prepared a briefing pack containing the physical items to be covered in the meeting, and a list of guide words and deviations to be applied to each item in turn. The objective of the meeting is to use each of the physical items as a focus for discussion, applying the guide words and deviations systematically to create lateral thought processes which reveal hazards in the minds of the experts. These are then documented by the HAZOP Secretary using standardised record sheets. Where information requirements exceed that available from the participants, actions are raised by the HAZOP Secretary for attention outside the session and reporting back to the facilitators.

Since the HAZOP technique is an intensive and therefore tiring exercise, it is important that short breaks are built into the meeting schedule to allow for frequent rest and refreshment. Experience has shown that attention and creativity diminish rapidly after about three hours of continuous meeting time, and it is therefore recommended that the maximum time between breaks is around two and a half hours. Continuous 8 hour sessions are to be avoided at all cost; in all probability such sessions will generate little useful output after the first three hours.

Another potential problem with HAZOP meetings is that of too many participants. In the quest for maximum expert coverage, large numbers of engineers and operatives are brought to the meeting. In addition to the two facilitators, a HAZOP team of approximately five is ideal, with a maximum of seven (5). This reference cites 'Shanahan's law', an anecdote stating that the length of a meeting is proportional to the square of the number of people present. The author is aware of one HAZOP meeting which took place with 40 participants, and has been asked to chair railway HAZOP meetings with a team of up to 18 persons.

Inevitably, personalities play a large part in the atmosphere of such large meetings. This leaves those having less inclination to dominate with little opportunity to contribute. Naturally, the HAZOP Chairman will attempt to control any HAZOP meeting in such a way that individual personalities do not dominate. However, his task is made increasingly difficult by having large numbers of persons in the meeting. There is, in the author's opinion, nothing such a large meeting can achieve that is not more effectively accomplished using separate focused sessions involving experts in each of the areas to be covered.

4.3   Consequence Ranking

The HAZOP record is a non-quantified listing of all potential hazards which apply to the railway system under consideration. Many of the hazards it contains will yield low risks once the consequences are taken into consideration. Examples of such hazards would be slips, trips and falls on wet station concourses (leaking roof panels). A method of ranking the complete HAZOP record to remove those entries which cannot yield significant risks is required.

It is convenient to sub-divide consequences into categories representing minor injuries, major injuries, single fatality and multiple fatalities. On this basis, each hazard can be allocated a consequence rating from one to four, and it is then a trivial task to group the HAZOP listing into four separate tables according to consequence alone. From this, a decision may be made on the level of risk quantification which may be required in the safety case, for example if only fatality-type accidents are to be assessed (categories three and four) or if major injuries are also to be included (category two).

4.4   Risk Assessment

There are two distinct methods of quantifying risks for the safety case. If sufficient operational experience exists using the same or similar railway equipment, for which accident data has been collected and recorded, then it will be possible to use this data to take the historical route to risk assessment. Generic data on railway accidents in the UK is published by the HSE in its annual Railway Safety Report series (6, 7, 8). It is likely that the use of data more than 5 years old will be overly pessimistic due to the improvements in railway technology and safety awareness that have taken place during this period. A good example of developing safety technology is the introduction of central locking on existing slam door rolling stock and the specification of driver-initiated plug type doors on new designs of passenger vehicles. Until only a few years ago, passengers falling from doors on moving trains was the largest single contributor to railway fatalities in the UK (9).

If insufficient data exists at the system level then a predictive approach will need to be employed. Failure logic modelling techniques such as fault trees and event trees may be used to model the interactions between safety-critical components and sub-systems for which failure rate data is available from published generic sources. The predictive approach has the advantage that it more properly handles the human factors element; human errors can be placed in the logic models in much the same way as component failures. Examples of a fault tree and an event tree for railway systems are illustrated in Figures 1 and 2 respectively.

4.5   Comparison with Risk Acceptability Criteria

Having calculated the separate risks due to each hazard, the total risk to a typical or most exposed person in each of the vulnerable groups must be assessed. It is this totalled value which will be compared with the risk acceptability criteria published by the approval authority.

Since not all risks will apply to each group, care is needed in selecting the defined groups to ensure that 'double counting' of risks does not occur. For example, staff may need to be separated into train-based, track staff and others to enable a logical summation of separate risk values to take place. A matrix has been found to be a valuable management aid in summating risks over a number of exposed groups and reduces the chance of accidental omissions. If a computer spreadsheet programme is available, the matrix can be constructed having exposed groups as the column headings with each identified hazard allocated a row to itself. Risk values can then be inserted into the cells as the calculation proceeds; the total risks are automatically displayed at the base of the matrix as the cells above are filled.

For the UK railway industry, risk acceptability criteria are published by Railtrack in the Railway Group Safety Plan (10). This provides safety targets for railway operators in the form of an accidental fatality risk for passengers per train journey, and a 'front line' track staff accidental fatality risk per year. It also gives major injury risk targets for these same groups which are lower ie. more onerous. Safety of the non-travelling public is addressed by the provision of a target accidental death risk expressed per year.

It is immediately apparent that the units of risk required depend upon the exposed group under consideration. For passengers (and train-based staff), an assessment of the number of journeys per annum will be required in order to link the assessed risk with the target.

Previous issues of Railtrack's own safety case have provided upper and lower level risk targets. The implication of these two-tier targets is that any group shown to have a total risk below the lower target will be considered to have an acceptable risk, with any group having a total risk above the upper target being considered to have an intolerably high risk. The region between these acceptable and intolerable zones is known as the 'As Low As Reasonably Practicable' (ALARP) region. In the ALARP region, risks can only be accepted if it can be demonstrated that all practicable measures to reduce these risks have been applied by the operator without prejudice to the service offered. Such arguments may involve the use of cost-benefit techniques.

5.   THE MANAGERIAL VIEW OF SAFETY

So far, the railway safety case has identified hazards, derived risks, and compared them with published acceptability criteria. It has made no case whatsoever that the safety standards claimed can be supported. This is the purpose of the safety management system (SMS).

Safety is a feature of organisational culture, akin to quality assurance, and not something which can be added once the railway operation is up and running. Indeed, the very management structure of the organisation itself should have been designed to ensure that the responsibilities and communications paths fundamental to safety exist. Safety awareness must percolate downwards to all levels from the top level safety policy statement endorsed by the Managing Director. An example of a safety policy statement is illustrated in Figure 3.

The SMS must be described in sufficient detail to give confidence that all safety claims can be achieved and maintained in practice, with the commercial pressures of a modern railway to contend with. The safety case must be able to convince the approval authority that safety amounts to more than a large amount of documentation covering procedures for working in a clean, relaxed, financially-unaccountable world. One way in which such a commitment can be demonstrated is by the mapping of safety management responsibilities directly onto sources of potential hazard. For example, in one train operating company with which the author has worked, the range of hazards arising from driver errors are monitored and controlled by the appointment of a Driving Safety and Standards Manager, in addition to the operational responsibility assumed by a separate Driver Manager.

A blame-free culture should be encouraged by senior management in which accident near-miss reporting is used to develop proactive safety measures to prevent major accidents from occurring. This culture will be reflected in the safety policy statement.

The safety management section of a railway safety case will ideally provide an organisation description, including a list of safety-critical post holders, safety targets to encourage continual improvements, communication paths for both normal and abnormal operations (including emergency response scenarios), and practical interfaces with other railway organisations and the outside world. This information is usefully supported by a senior management organogram and flow charts showing safety-related information flows during both normal and abnormal conditions.

6.   SUMMING UP STATEMENTS

The summing up statement is expected to draw together the principal pieces of evidence presented in the earlier sections of the safety case, showing that the railway organisation is a suitable and competent operator and therefore worthy of a certificate to operate. It is helpful to think of this final section as the speech which might be given by a prosecution or defence counsel in the courtroom just before the jury retires to consider its verdict.

Reference should be made to important findings of the hazard identification and risk assessment work, showing how the organisation has put in place pro-active safety measures to prevent those hazards becoming accidents. It is not a failure to identify hazards in the proposed operation, but it is a failure to do something about them.

The summing up statement will show that the SMS has been structured in such a way that safety levels can be set, and most importantly, maintained at an acceptable level throughout the lifetime of the service, and that regular audits and safety inspections have been appropriately resourced to highlight developing deficiencies.

It is essential that the summing up statement is clear and effectively worded. Ambiguity cannot be afforded at this stage; if it is allowed to occur then doubt may be thrown over the rest of the safety case.

7. THE SUBMISSION PROCESS AND BEYOND

In the UK, Railtrack is the approval authority for railway operator safety cases. Whilst being flexible over the exact structure and detail of a safety case, several companies with whom the author has worked have found it most helpful to explore with Railtrack what their individual requirements will be prior to submitting a draft document. Exploratory talks have been held between senior safety managers of both organisations and from this a safety case format and level of detail has been agreed. Meetings of this nature also allow key Railtrack personnel to gather confidence in the operator's commitment to the safety case process. Time invested here can only help to improve the operator's chances of a successful outcome.

Besides preparing the ground for a future submission, this process has allowed companies to budget far more accurately for the man effort, timescales and costs involved in preparing their documents. It has also highlighted the need for specialist safety assistance early in the process so that a suitably-qualified consultant can be appointed to the safety case team.

Award of the certificate to operate is not the end of the safety process. It is merely the beginning of the next step. The Railways (Safety Case) Regulations (1) state clearly that if any changes are made to the organisation which may have an effect on safety, then the safety case must be updated accordingly and re-submitted. Examples of such changes would be the introduction of a different class of T&RS, a change of route, or a significant change to the facilities offered to passengers at a station (for example de-staffing).

If the safety case has been structured logically, then making changes will be a relatively simple exercise, and appropriate amendments can be forwarded to Railtrack without the need to re-submit the entire document. In turn, the authority will be able to deal efficiently with the proposed changes and provide its approval to continue operations. As soon as Railtrack has provided its approval, the document control system within the organisation can be brought into action to circulate the amendment to all copies of the safety case.

8. CONCLUSION

This paper has addressed some of the major issues involved in planning, constructing, submitting and maintaining a railway safety case. It has offered practical advice which can be applied by a variety of organisations to assist in the whole life cycle of their railway safety cases.

It has illustrated important points by reference to real examples, where lessons have been learned under difficult conditions leading to delays in the award of certificates to operate. The need for logical structure has been stressed, and a safety case format has been suggested which, with a minimum of adaptation, should be suitable for most needs.

One of the most significant factors to emerge is that the safety case need not be a complex document. To some extent its complexity will reflect the internal complexity of the organisation which has prepared it. Since simplicity is one of the key factors in minimising communication errors within any organisation, it must also be true that simplicity is key to achieving safety. Above all else, the railway safety case must be shown to be clear, credible, and consistent throughout.

REFERENCES

1   Railways (Safety Case) Regulations 1994, (Guidance on Regulations), February 1994, Health and Safety Executive Doc. No. L52.

2   Railways (Safety Critical Work) Regulations 1994 - Guidance on Regulations, February 1994, HSE Publication L50.

3   Carriage of Dangerous Goods by Rail Regulations 1994 - Guidance on Regulations, April 1994, HSE Publication L51.

4   Duxbury, H. A. and Turney, R. D. Techniques for the analysis and assessment of hazards in the process industries, Seminar on safety and hazards evaluation, New Mexico Technical Research Centre for Energetic Materials, 1989, pp 1-14.

5   Swann, C. D. and Preston, M. L. Twenty-five years of HAZOPs, J. Loss Prev. Process Ind., 1995, Vol. 8, No. 6, pp 349-353.

6   HM Railway Inspectorate's Annual Report on the Safety Record of the Railways in Great Britain during 1992/1993, December 1993, HSE/HMSO.

7   HM Railway Inspectorate's Annual Report on the Safety Record of the Railways in Great Britain during 1993/1994, December 1994, HSE/HMSO.

8   HM Chief Inspecting Officer of Railways' Annual Report on the Safety Record of the Railways in Great Britain during 1994/1995, November 1995, HSE Books.

9   Passenger Falls from Train Doors - Report of an HSE Investigation, March 1993, HSE/HMSO Publication C60.

10   Railway Group Safety Plan 1995/96, Safety and Standards Directorate, Railtrack plc.

DISCLAIMER

Whilst supporting the views expressed in this paper, the author makes no claim that following these guidelines will guarantee acceptance of a railway safety case by any authority.

ACKNOWLEDGEMENT

The author wishes to thank the management of EQE International Limited for permission to publish this paper.

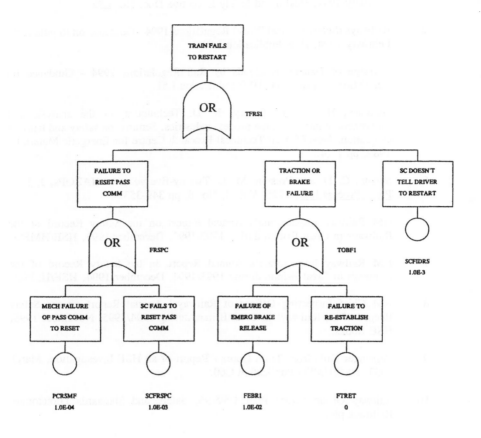

Fig 1 Example fault tree for failure to restart train following passenger emergency

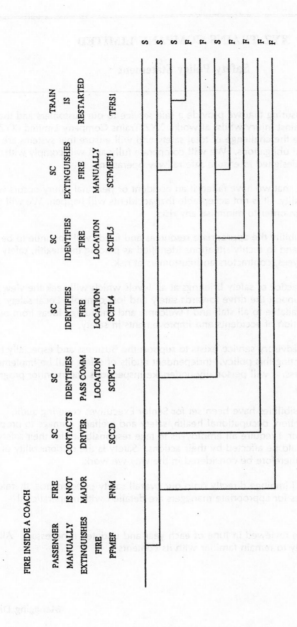

Fig 2 Example event tree for train stopped with fire inside a passenger coach

XYZ TRAINS COMPANY LIMITED

Safety Policy Statement

I am committed to ensuring that we provide a safe service to our customers and that our staff are protected against injury whilst at work. XYZ Trains Company Limited (XYZT) will vigorously pursue the challenge of total safety and will ensure that systems are in place to meet all legal obligations. We will co-operate fully and will comply with Railtrack procedures designed to ensure safe railway operation.

Our approach will be that we have failed if an accident or personal injury occurs within our area of responsibility. It is not acceptable that accidents will happen. We will assess all potential hazards in order to minimise any risks.

Within XYZT responsibility the appropriate resources and funds will continue to be made available for training and to rectify situations identified as placing the health, safety and welfare of our employees, contractors and customers at risk.

I will continue the practice of safety briefings at all levels which will seek the views of our employees and support the drive towards safety and loss control. Local safety representatives are available to all staff and I welcome and encourage ideas from our employees for prevention of accidents and improvements in safety.

A professional safety advisory service exists to support the Business and especially to assist me in implementing this policy. Independent audits of safety will be implemented throughout the Business. I will personally undertake inspections and monitor progress against objectives set.

Specific safety responsibilities have been set for Senior Executives covering audit, legislation, fire prevention, occupational health, safety and welfare, damage to property and security. However, I require all employees to take responsibility for their safety and that of others who could be affected by their actions. Safety is the responsibility of all employees and must therefore be considered in the way we work.

Operational safety will impinge directly onto our overall safety performance, therefore specific responsibilities for appropriate managers are detailed within operational procedures.

This document will be reviewed in June of each year and updated if necessary. All employees have a duty to remain familiar with its contents.

1 January 1996   **Managing Director**

Fig 3 Example safety policy statement

C511/11/052/96

A survey of train to track CCTV transmission systems

S FARAROOY BSc, MSc, PhD, CEng, MIEE, MIEEE
School of Electronic and Electrical Engineering, University of Birmingham, UK
J ALLAN BEng, PhD, CEng, MIEE, MIRSE
School of Electronic and Electrical Engineering, University of Birmingham, UK and also
London Underground Limited, UK
A EDWARDS
Engineering Directorate, London Underground Limited, UK

SYNOPSIS

The paper describes a study carried out into state-of-the-art technology for transmitting platform camera video pictures into train cabs. The objectives of the study were to review the market place, consider the benefits/disadvantages of the systems available and to recommend a system to be installed, on a trial basis, using enough equipment to evaluate the effectiveness of the type of equipment chosen.

1. INTRODUCTION

London Underground Limited (LUL) has been engaged in a project, known as One Person Operation (OPO) enhancements, to provide monitors and mirrors on platforms in suitable housings to enable the train operator to view the platform/train door interface. This enables the person operating the train (the driver) to close the train doors and proceed from station stops in a safe mode, ensuring nothing or no-one is trapped in the doors or in the gaps between the train and the platform before moving away from the platform. The current platform based system installed incorporates a number of design improvements over previous equipment used and has enhanced both the system availability and passenger safety.

The existing system, however, has a number of limitations. The main one is that the view of the platform/train door interface is available only whilst the train is at rest or in its first few feet of forward movement. This is the time of greatest risk, but it is possible for incidents to occur at other times before the train arrives at the platform or until the rear of the train leaves the platform. Another disadvantage is that the driver cab has to stop in a position which enables a complete view of the platform mirror/monitors. There are other problems of positioning and maintainance of platform-based system as well as the consistency in the quality of pictures.

LUL's ultimate aim has always been to install Platform to Train CCTV Transmission Systems for its OPO scheme in order to address some of the above limitations; but the biggest constraint has been related to the difficulty involved in locating additional equipment into the train cab in existing rolling-stock. Such systems have so far only been pursued when major rolling stock replacement programmes are undertaken, for example on Central Line, more recently the Jubilee Line Extension (JLE) and the Northern Line renewal projects. Refurbishment of existing rolling stock gives an opportunity to review the current platform based system and to install the best available technology for in-cab CCTV.

The study followed a systems engineering approach [1] by starting with the evaluation of LUL's operational needs in order to be able to formulate and produce a consistent user requirement specification for the Platform to Train CCTV Transmission Subsystem.

This included interviews with Piccadilly Line's Train Services Manager, Central/Northern and JLE project team members involved in specifying such systems as well as Piccadilly and Central Line drivers. The most important aspect to be decided was the distance of coverage of the path of train before, during and after its stop at each platform (outdoor, tunnel, straight and curved) by the transmission system. Next came the question of performance and life cycle costs which may include an important maintenance cost element.

Review of currently available systems commenced with a visit to the Far East and interviews with railway operators as well as manufacturers of such systems. This was continued with visits, meetings or written communication with European and UK manufacturers, suppliers and operators of in-cab CCTV systems. This aspect of the study concentrated on the evaluation of feasibility of alternative infrared/microwave solutions for LUL in comparison with the radio-frequency radiating cable system in operation on the Central Line since 1989.

The outline of this paper is as follows. Firstly, in the next section, the operational need of the in-cab CCTV system is established based on interviews with LUL train services managers, engineers and train operators/drivers. In section 3, various CCTV free space transmission technologies are briefly outlined. Systems in operation or under trial in the Far East are reported in section 4. Section 5 compares the systems in use (Central Line) or specified for use (Jubilee Line Extension - JLE and new Northern Line) within the LUL. In sections 6 and 7, products and services of system manufacturers and suppliers in Europe and the UK are summarised. System requirements specifications for platform to train CCTV transmission sub-system are discussed in section 8. A comparison of various transmission technologies is then summarised in section 9. Conclusions and recommendations are presented in section 10. References and acknowledgements are provided in the final sections 11 and 12.

2. OPERATIONAL NEEDS

In-cab CCTV system provides a number of advantages over platform-based mirrors and monitors which relate to the overcoming of the following shortcomings:
1. Reflective light at certain times of day/seasons (Oct.-Nov. and Feb.-March) due to the low angle of sun with respect to the horizon; and reflection from walls, buildings and trains;
2. The need for daily maintenance by station foreman as well as physical end of platform limitations/crowding;
3. Environment: rain on windscreen, graffiti, damp, etc. of platform monitors;
4. Limited Drivers' view: Some platforms are only train length and not sufficient room for a good view because of crowding at the platform ends; Drivers' view is also limited due to point 3 above.

LUL have carried out Quantitative Risk Assessment (QRA) studies into the risks of boarding and alighting of the trains by passengers. These have pointed towards the use of in-cab CCTV as an important contribution to substantial increase in passenger safety.

Interviews with train drivers generally pointed to the operators' requirement of observing platform activity before entering platforms. In the stations where this view is limited for the train operator (tunnels in general and curved ones in particular), the in-cab CCTV should cater for this requirement. The possibility of switching the CCTV off between platform entrance and the stopping point is also generally ruled out. In other words, the findings confirm the requirement for continuous CCTV pictures 50 metres from platform entrance until the rear of train leaves the platform (a total distance of 250 metres on average) for underground stations.

Three more observations were made: i) The need for test locations on the in-cab system (required for example to record the quality of pictures onto a video recorder for test purposes); ii) Requirements for continuous picture to be available beyond the points and crossings; iii) the need for a sharp cut-off and hold-off of CCTV pictures at the end of the transmitting range.

3. PLATFORM TO TRAIN CCTV TRANSMISSION TECHNOLOGY

Methods of transmitting a video signal from a camera to a monitor in a cab include transmission technologies employing radio, microwave or infrared frequencies which enable dynamic as well as point to point transmission so long as there is line of sight (LOS) between the transmitter and receiver. Other parameters for consideration include the range of transmission, performance and quality, security and reliability and also costs.

In terms of the transmission frequency, it is important to note that all these techniques relate to the use of waves from the electromagnetic spectrum. The bandwidth required for the transmission of one channel of video signal depends on the resolution (For every 100 lines of resolution, a bandwidth of 1 MHz is required). The chrominance signal is modulated on a 4.43 MHz carrier wave in the Phase Alternate Line (PAL) system, therefore a colour signal has typically a bandwidth of at least 5.5 MHz.

4. CCTV TRANSMISSION SYSTEMS IN OPERATION IN THE FAR EAST

Table 1 summarises the infrared (IR), microwave (MW) and radio frequency (RF) based CCTV transmission systems currently in operation or under trials in the Far East as determined from this study. Here, the Hitachi's Infrared system is further elaborated upon.

Hitachi's Infrared (IR) LED system has been in operation in the Sendai Subway system since 1987 after trials as early as 1982-83 [3,4]. This system has employed black and white pictures but has been updated to provide two way communication including downloading of train information at the maintenance depots. Since then four other railway and subway operators have been or are implementing Hitachi's IR/LED system in their drive to one person operation. Two of these were visited, namely the Osaka Municipal Transport Bureau (OMTB) and Tokyo Municipal Government Subway (TMGB). The OMTB have been involved in extensive trials (since 1990) of the IR system for the new Tsumuri Ryokuchi line and will start operating the system from Oct. 1995 after all trains have been equipped with the system and all the operators have been trained in their usage. The system is analogue and FM modulation is used. The monitors are 10.3" LCD type from Hitachi. The trains are 4 car set total length of 60 metres. Simulations have been carried out for curved tunnels and 4 transmitters can cover an effective area of 70 metres. Receivers are situated in the back of the train so that the transmitters can be installed inside tunnel near to the platform and cover the whole of the operating range, i.e. the distance from 5 metres before the train stopping point at the station until the rear of the train leaves the platform. Straight platforms require a minimum of 2 transmitters. Tokyo Municipal Government Subway (TMGB) have employed Hitachi's infrared system on their modern 12th Line (Hikarigaoka). System's specifications include: 850nm near infrared, bandwidth of 4.5 MHz, train to trackside transmission at 1 Mbps, FM modulation at 19 MHz carrier frequency, video resolution of 525 lines (NTSC), 30 frames/sec, 10.4" LCD monitor, one split-screen picture per channel. Three infrared transmission channels are provided to avoid problems of cross-talk.

Operator	Transmission media	Supplier	Status
Singapore Mass Rapid Transit Corporation	Microwave@23.5GHz	Thompson/Solitech	In operation since 1985/6
Hong Kong Mass Transit Railway Co. (MTRC)	RF+Radiating Cable	Siemens	Trial Phase1 completed Phase2 starts soon
Hong Kong MTRC	Microwave@2.4GHz region (3 channels)	Philips/PELCO	Trial completed
Hong Kong MTRC	Microwave Wave Guide @2.4GHz	GEC-Alsthom Transport St.-Oven Cedex, France	part of an integrated ATO/ATC system
Hong Kong MTRC-Airport Railway	Infra-red Class 1 laser	GPT-Plessey=Contractor Lee Communications	System specified
Japan-Sendai Subway	Infra-red LED (800-850nm)	Hitachi	In operation since 1987 Trials 1982-83
Tokyo Municipal Government-12th Line	Infra-red LED (800-850nm)	Hitachi	In operation since 1992
Osaka Municipal Transport Bureau	Infra-red LED (800-850nm)	Hitachi	Trials since 1994, operation from Oct. 1995
East Japan Railway Co. Shinkansen Tokohu Line	Microwave@45GHz	Hitachi	Trials since 1993; Major introduction 1997
Tokyu (Private) Railways	Microwave@45GHz (4 video channels)	NEC	Trials since 1994/5

Table 1- Brief Summary of Systems in Operation/under Trial in the Far East

5. CCTV TRANSMISSION SYSTEMS IN OPERATION WITHIN LUL

CCTV transmission systems within LUL in operation (Central Line - CL) or specified for implementation (Jubilee Line Extension - JLE and new Northern Line - NNL) are all based on Radio-Frequency and radiating cable technology. Central Line's in-cab CCTV system has been a welcome development for train operators. A similar RF solution has been adopted for JLE and NNL but with further improvements in mind. Both the new systems provide additional ruggedness to track maintenance activities and coverage over points and crossings. They propose to install the leaky feeder (or its new tube radiating antenna RTA replacement) sideways under the platform nosing; and to fit two sets of receiving antennas in both front and rear of the train and switching from one to the other. This approach increases the system complexity and introduces the need for video transmission across the autocoupler.

6. SYSTEM MANUFACTURERS IN EUROPE

A number of European manufacturers were visited or communicated with. Most important products included Grundig's infra-red, Siemens's Radio Frequency (SITRAIL) and Infra-red systems, Velec's RF solution using AM modulation and GEC-Alsthom's IAGO Microwave guide system. These are introduced in more detail in this section.

Grundig Electronics have developed an infrared LED system. The special features of their system include employing a horizontal array of 12 LED's for transmission of IR beams (to increase horizontal angle of coverage) as well as an automatic beam finder (servo system) on the receiver which locks onto the beam to ensure that any movement of the train in the vertical plane as it approaches or leaves the station does not cause the loss of signals. The system has already passed relevant EMC tests of the German telecommunications authority (PTT). Unlike other optical systems (e.g. fibre-optics) which use high intensity laser beams, the IR beams present no hazards to system installers, track-side (permanent way) or rolling-stock staff. The system has a wavelength of 850nm and a bandwidth of 5.5MHz for one channel (full split-screen) of communications. Time multiplexing will allow 4 channels to be used but each at 12.5 Hz (12/13 frames/sec.). The video transmission is analogue using FM modulation and has a resolution of 625 Lines of colour. The coverage area for each transmitter is 135 metres. The transmitter position will depend on the location of the receiver on the train and will be at the same height. The power supply is inside transmitter housing and very compact, only the length of transmitter has to be long enough to enable coverage by the lenses. Multiple transmitters may be installed at intervals and the signal can be 'cascaded' as train moves from the range of one transmitter to another. The signals can be coded to avoid problems of cross-talk. The system has been in trial in Germany and by the Cross Rail Project.

GEC-Alsthom's microwave (2.3-2.4 GHz) waveguide system (IAGO) is a very high bandwidth (200 MHz) data, audio and video transmission system which can be used for bi-directional track-train transmission, train to train communication as well as train localisation and potentially signalling (ATP/ATO). IAGO [6] stands for *Informatisation et Automatisation per Guide d'Ondes* (Computerisation and Automation through a leaky Waveguide). The concept has been developed since 1987 and railway validated (from 1988-1991) by the French subway operators (RATP). Trials of the system have been carried out by Lyons Subway (1991) and French National Railways, SNCF-Freights (1993). The 2.3-2.4 GHz band is limited for use in France for military usage, but the 2.4-2.5 band is available under UIT standards for Scientific and Medical Instruments (SMI band). At such frequencies, there is a low level of attenuation which compares favourably with radiating coaxial cables operating at far lower frequencies.

The microwave waveguide is a hollow (2 mm thickness) rectangular electric conductor (Aluminium) of 5 cm height and 10 cm width. At intervals of 6 cm, narrow slots are stamped/punched along its top surface so that at the range of microwave carrier frequencies, a uniform field is radiated so that a receiver situated at a distance of at least 30 cm can pick-up the signals. The waveguide is protected against water and damage by an adhesive film and a layer of Glass Reinforced Plastic (GRP). It is therefore very rugged and suitable for railway environment conditions. Audio and video analogue signals are FM modulated, whereas data is transmitted digitally (PSK modulation) at 2 Mbps. The waveguide transmitter may typically be laid in the 4ft or just on the side of one of the running rails. The receiving antenna is situated under the car body vertically above the transmitter at a typical distance of 30 cm. The transmitting equipment may be situated at a maximum distance of 10 metres from the transmitter fed from standard video signal from the CER.

Siemens has a number of SITRAIL radio frequency (RF) and radiating cable installations in many countries including LUL's Central Line, and has recently developed and carried out trials of an infra-red (IR) system in Karlsruhe, Germany. SITRAIL (Siemens Television equipment for RAILway application) was developed in close co-operation between the Hamburger Hochbahn and Siemens. The system consists of a half-picture divider (adding pictures from two cameras into a split-screen picture), FM modulator, demodulator, transmitting cable, receiving antenna and an in-cab monitor. Two types of transmitting antennas have been developed: i) a radiating/leaky coaxial cable, the type of cable normally used for installations in tunnels; and ii) a Radiating two-conductor metal Tube/Aluminium bar line Antenna (RTA), which is robust enough to withstand the mechanical stress caused by the vibration of sleepers on the ballast bed.

The Siemens IR systems uses light emitting diodes (LEDs) rather than laser diodes. The range of the transmitter depends on the transmission angle. In sharp bends of 25 metre radius with panoramic illumination, the range is 20-30 metres. In larger and more straight stations, this can extend to 120-150 metres. The overall system is designed such that 4 to 6 transmitters can be connected to the basic unit. Trials of this system has recently been carried out in Germany on a tramway platform with sharp curvatures (Hochstetten Kreis Karlsruhe).

Velec has been working for more than 20 years with French railways (SNCF) and metro operators (RATP) and rolling stock manufacturers to develop and industrialise reliable and high performance electronic equipment suitable for railway environment. Velec's platform to train radio/CCTV transmission system is based on RF transmission with radiating cables. Its black and white version has been operational by SNCF for more than 15 years. The colour version, to be installed on LUL's Jubilee Line Extension, consists of 236 LCD colour monitors located in 118 driver cabins. Overall, 39 platforms will be equipped with 2 or 4 cameras and transmitting equipment including radiating cable installed under platform nosing. The colour system technical characteristics are: Single video band Amplitude Modulation (AM), central frequencies of 50.2 and 62.86 MHz, video bandwidth of 5 MHz, 5.7" LCD monitors (larger 8.6" monitors are also available), CCD standard colour cameras with Genlock facilities, Input video signal of 1 volt peak to peak (75 ohm TV standard CCIR, PAL). Power supply requirement for trainborne equipment is 52 volts DC.

7. SYSTEM MANUFACTURERS AND SUPPLIERS WITHIN UK

Two UK manufacturers of laser systems have been investigated whose products are introduced here.

Lee Communication Limited specialise in the design and manufacture of optical free space transmission systems. Their latest product, Transport Rail Approach Communication Systems (TRACS), provides video, audio and data transmissions from the station platform to the oncoming train. It can provide up to 4 channels of real time outputs and can be configured to offer multi-channel audio for applications such as station to train PA announcements. The fundamental transmission medium consists of a multiple beam infrared laser source with a multi-cell photo diode receiver. The prototype has a single laser source and a single lens receiver. The source is intensely modulated with four FM video carriers in the range of 60 to 140 MHz. These channels have a nominal FM deviation of ±8 MHz, and the video signal is pre-emphasised to the CCIR 405-1 standard. Deviation limiters are fitted, and the overall baseband video signal bandwidth extends beyond 10 MHz. Therefore sound or data subcarriers could be used on every video baseband channel.

Vector Technology Limited has been in the field of low power laser and laser technology since 1989. It has recently developed an optically linked train surveillance system, each link carries 3 real time channels for audio, data and video. The audio and data runs through the train in a continuous loop, so that each carriage receives continuous audio and data originating from the central carriage. Video pictures from on-train surveillance cameras are fed to the driver's cab. The system may be used for in-train video transmission, say from the

© IMechE 1996 C511/11/052

receivers on the back of the train to the cab in front, thereby overcoming the need for video transmitting cable connections through the autocoupler.

8. SYSTEM PERFORMANCE SPECIFICATION

The system performance specification were developed on the basis of existing LUL standards and specifications, system standard for OPO Track-to-Train CCTV, system standard for CCTV systems, system standard for OPO monitors and mirrors and relevant JLE specifications. A number of modifications were suggested, firstly, to encompass infrared and microwave solutions as well as the current radio frequency and radiating cable one. Secondly, based on a re-examination of operational needs and interviews with operators, the range of coverage of the transmitters have been extended to include 50 metre approach to the underground platform entrance.

9. COMPARISON OF CCTV TRANSMISSION TECHNOLOGIES

A summary of the advantages and potential problem areas for different technologies (media) for track-to-train CCTV transmission is given in the table 4 below. Problems are constantly being tackled during new trials of various systems by the operators in close association with system manufacturers. There, however, always exists a trade-off between a number of conflicting performance aspects and a solution is therefore dependent on the requirements of each particular application.

Technology	Advantages	Potential Problems
Infra-red	• Tried & Tested: Hitachi in operation at Sendai since 1987 and by 4 other railway operators in Japan • No license requirement • No Interference problems • Large Bandwidth	• May suffer from optical noise and blocking • Uses lenses: Staining requires cleaning and maintenance • Narrow Beam: Alignment problems
Microwave	• Tried & Tested: Thomspon's in Singapore (static/short range) • Large Bandwidth: Up to 4 channels possible if required • Twice the range of infra-red	• Interference from poles and multipath in tunnels • Cross talk problems due to wide angle of transmission • License requirement
RF+Radiating Cable	• Tried & Tested: Siemens in LUL and Hamburger Hochbahn • License not problematic in UK	• Installation (particularly over points & crossings) and maintainability problems for leaky feeder • License requirement • Small signal/noise (CL): requires amplifier to achieve good quality pictures

Table 4: Comparison of CCTV Transmission Technologies

10. CONCLUSIONS

In this paper, the operational need for platform to train CCTV transmission was re-examined. Based on interviews with train operators, it was concluded that the requirement for the availability of platform pictures in-cab from a distance of approximately 50 metres from station entrance was desirable for underground stations, in particular those with curved platforms. Various CCTV transmission technologies in free space, namely, radio-frequency, microwave and infrared, were introduced. A comprehensive survey of currently available systems in the market place was presented. Railway operators' experiences of some systems in operation in the Far East, Europe and UK were noted. Products from system manufacturers in Japan, Hong Kong, Europe and the UK were briefly described. In-cab CCTV systems in operation within LUL's Central Line and those specified for implementation on the new Northern Line and Jubilee Line Extension projects were compared. Comparison of alternative technologies in terms of the use of different media for CCTV transmission was summarised. The appropriateness of any solution depends on the considerations for each particular application.

The performance and cost of each solution would therefore have to be judged against an agreed and consistent set of system requirements specifications. Two systems were recommended for trials by LUL. One, infrared solution, is suitable for refurbished rolling stock as originally specified in the scope of work and another, IAGO's microwave waveguide system, is believed to be suitable for consideration in future LUL mega projects - renewal, extension or new lines.

11. REFERENCES

1. ALLAN, J and J S WILLIAMS. Systems Engineering Code of Practice for Maga Projects, *IRSE Proceedings*, 1995/96.
2. CONSTANT, M and P TURNBALL. The Principles and Practice of CCTV, *Paramount Publishing*, 1994.
3. JITSUMA, U, H OSHIMA, H AKIYAMA, T ARAI and R TASHIRO. Real-time video signal transmission system from wayside to moving trains using infrared beam as carrier, 1984.
4. OSHIMA H, T ARAI and S KAWAHATA. Computer Control Systems for the Sendai Subway, *Hitachi Review*, 1988, **37**-6, pp 385-392.
5. KUSAKARI M, M SAKUMA, E ISOBE and T TAKAOKA. On-board Train Information Control Network Systems, *Hitachi Review*, 1991, **40**-4, pp 303-308.
6. RAVEN, K.L. and M. STACH. Drahtlose Zugfernsehenlage der Metro Amsterdam, *Siemens-Energiotechnik*, 1979, 1-5, pp 171-175.
7. DUHOT, D, M HEDDABUT and P DEGAUQUE. IAGO: Transmission on Radiating Waveguides in the Transport Field, *GEC-ALSTHOM Technical Review*, 1991, No. 6.

12. ACKNOWLEDGEMENT

The authors wish to express their gratitude to London Underground Limited for its permission to publish this paper.

C511/11/050/96

Automatic train protection and the operational railway

B D HEARD BSc, FIEE, FIRSE
British Rail Projects, London, UK

SYNOPSIS

The paper defines Automatic Train Protection (ATP) and Cab Signalling. It reviews the recent history of development of both, and describes the effects of both continuous and intermittent systems on capacity. The risks of installation and maintenance of lineside-orientated systems are highlighted, along with the benefits of removing the infrastructure to track-remote locations. Finally, the emergence of the European Train Control System (ETCS) is explained, together with the many advantages it offers.

1. INTRODUCTION

It is worth defining the various terms associated with modern on-board train control systems.

1.1 Automatic Train Protection

is employed to ensure that a train never travels at an instantaneous speed greater than that which is safe. The safe instantaneous speed may be dictated by the maximum permitted train speed, the maximum permitted line speed, the presence of a permanent or temporary speed restriction, or the speed while braking towards a restriction or stopping point.

1.2 Cab Signalling

provides information and instruction to the driver in the cab rather than in the traditional manner, at the lineside. Systems generally advise the maximum permitted instantaneous speed, the target speed and sometimes the target distance. Some earlier systems have been tried which would better be called "Aspect Repeating" since they provided in-cab replicas of the lineside signal aspects. However the author is not aware of any of these still in use.

It should be noted that ATP and Cab Signalling can each exist independently of the other. It is quite possible to provide an ATP system which gives no indication of its status whatever to the driver. A simple example of this is the London Underground train stop system, in which the driver's first advice of its intervention is when the train pipe is totally and rapidly vented. Equally, one could provide a perfectly adequate cab signalling system with no interface to the train braking or power control systems. Such a system is unlikely to be adopted, though, since the incremental cost of providing an intervention curve and the interfaces would not be great.

Many systems recently introduced, or being planned, however, combine the two, in some cases superimposed on existing infrastructure, complete with its lineside signals, and in others with new or heavily modified infrastructure, eliminating the lineside signals.

2. OPERATIONAL CHARACTERISTICS

There is no doubt that train safety is increased by provision of ATP, but the question of its effect on operating performance is quite complicated. Some consider that its introduction must allow trains to approach each other more closely, thus reducing headway times, or to approach stopping points faster, thus benefiting from a higher speed once the stopping point is removed (i.e. a signal is cleared). Both of these may be true with certain types of system, but in the cases of many systems currently in use, the opposite is true.

There are two basic ways in which the data needed by the train can be provided. The first, known as continuous usually operates by antennae on the train passing over a radiating source throughout the journey. The second is known as intermittent, and uses some kind of beacon placed in or by the track at chosen points. A detector on the train reads these beacons and thus up-dates the on-board data.

The choice between these two is vitally important when planning a project; whereas the continuous system has the potential to upgrade operating performance, this has until now, usually been at a substantially greater cost than the introduction of an intermittent system. The adoption of an intermittent system, though cheaper, does not generally improve operating performance; indeed , it can worsen it. Figure 1 explains this. Assuming a train requires 2000m from 100m/h to stop, curve (a) represents the calculated service braking curve for this performance. An ATP system would impose an intervention curve typically as shown at (b), about 6 m/h above the service curve. Today, (without ATP) drivers tend to perform according to curve (c), by braking quite heavily on first warning of the stopping point, gradually releasing the brakes once they have halved the speed, and preparing to brake very gently to a halt if the signal has still not cleared. If the signal clears when the train is 1000m from it, 3 very different scenaria are created according to whether continuous, intermittent or no ATP is present.

If continuous ATP is in use, the on-board equipment will detect the removal of the stopping point more or less instantly, and the intervention curve reverts to 106 (say) m/h. The driver can accelerate and will be able to proceed according to curve (d).

If intermittent ATP (at its simplest) is in use, the train will need to continue braking until it arrives at the beacon by the signal, since only then will the on-board data be up-dated. Thus the train will be virtually at a stand before release is given. (In practise an additional beacon is normally placed about 300-400 yards before the signal so as to give an early release if it has cleared).

If no ATP is fitted, assuming the driver can see the signal, he will begin to accelerate as soon as he sees it clear, and his train will follow curve (e).

This still over-simplifies the arguments since it presumes drivers with the benefit of continuous ATP will follow the service braking curve; that no additional beacons are fitted in the case of Intermittent ATP; and that all or most drivers really do follow curve (c) without ATP, and cannot be trained to drive closer to curve (a). Even so, there is clear evidence that continuous ATP can improve headway performance, and Intermittent ATP can worsen it.

3. TRACK TO TRAIN TRANSMISSION

As stated above, various methods have been used. In the case of continuous transmission, either the rails are used, by adding codes to the track circuits (which are there to detect the presence of a train on a particular section of track), or by placing a conductor– a light section cable – along the centre of the 4-foot, and injecting codes into it. Both methods have disadvantages. The type of railway likely to be equipped with ATP is also likely to have continuously welded rail, and the cost of inserting insulation at the conjunction of two track circuits is high, both in money and reliability. Therefore electrical separation (by means of tuned circuits) is almost universally adopted, but this usually means the code signal strength is quite low in the area of a joint, so that a slow-moving train (for example, one just starting after a signal stop) may lose the code long enough for the on-board equipment to decide a fault has occurred. Also, it is necessary to ensure the codes are injected ahead of the train, so that the leading antennae can read them. This means either that the line cannot be used for bi-directional operation, or that an expensive, relatively complex method of transposing the two ends of the track circuit is fitted. With or without coding for ATP, signal engineers will admit that the track circuit is not the most reliable of mankind's inventions. It is subject to many influences outside immediate control - weather conditions, vandals, permanent way staff etc. In the event of its failure, therefore, the ATP codes are lost, as well as that section for the train detection system.

The conductor in the 4-foot avoids most of the problems of combining ATP coding with the track circuit, but is, if anything, even more vulnerable to damage. However, at least one, quite advanced, system, divides the conductor into 300m sections, and has intelligence sufficient to "leap" one such section should it have failed. The system, though, is not a low cost one, but it does provide high operational reliability.

Turning to intermittent systems, these all use some form of beacon mounted in the 4-foot or immediately outside it (e.g. on the sleeper ends). The beacons take one of two forms: those which radiate all the time, power being provided from the lineside, and those which are "awakened" by a message from a passing train, and only radiate while the train antenna is immediately above them. (The train-to-track message also provides the power needed for the beacon to radiate.) The latter are generally referred to as passive radiators, and systems using this type have the advantage that a beacon can quickly be installed to warn of an emergency speed restriction. Bi-directional operation is not a serious problem for intermittent systems, since either they can be offset from the centre line of the 4-foot, so that only traffic travelling in the relevant direction reads them, or they can be placed on the centre line and suitably encoded according to the direction of the signalled move. All types of beacon, though, being physical objects mounted in or about the 4-foot, are at risk of damage.

Generally, all types of transmission medium not associated with track circuits offer higher reliability than coded track circuits, but nevertheless, this is never 100%.

4. INSTALLATION AND MAINTENANCE

It is axiomatic to state that staff are required to work in or very close to the 4-foot both to install and maintain the equipment used to provide the track-to-train communication systems discussed above. It is equally axiomatic to state that such staff are at considerably greater personal risk than when they are working in, say, an equipment room. There is, therefore, a great advantage in producing systems which minimise the amount of track mounted equipment, and move it at least to the lineside, or better, into equipment rooms which might be quite remote from the operational railway.

At present, one has to build into the overall cost of developing, installing, testing, commissioning and maintaining an ATP system, the probability of staff being injured, or even killed during each activity. There is little point in providing ATP on a route, expected to reduce passenger deaths/injuries by 5% if it increases the probability of death/injury to staff by a similar amount.

A way around this problem is beginning to emerge, but first it is worth looking at the requirements concerning integrity of ATP systems, or, to use what is now becoming "old-speak", the fail safe requirements.

5. LEGISLATION

Ever since the first Englishman was blessed with the title "Signal Engineer" (late 1800's) the fight has been to achieve Absolute Safety. In other words, no matter what item of equipment failed, or whatever its mode of failure, the system would revert to a more restrictive condition rather than a less restrictive one. For example, if a semaphore signal wire broke, the arm was weighted so that it went to the "Danger" position. Track circuits are designed so that any component failure will cause Track Occupied rather than Track Clear to be indicated.

However, in recent years, as systems become more and more complex, and as Software becomes more and more part of the systems, the Signal Engineer has come to realise that the pursuit of absolute safety is akin to banging one's head on a wall. To some extent, we have to thank the aviation and nuclear industries for bringing this to our notice, since they, very soon after their emergence, realized that Absolute Safety is an unachievable goal.

This is not to say that Signal Engineers relaxed their opinion on the need for safety. Instead, they have come to accept that it is calculable, and systems can be analysed, even before they are constructed, to determine the likelihood of a dangerous situation arising. There is a learning curve associated with this change of thinking. For example, in setting improbabilities using exponents of 10, it is not too difficult to require times before a dangerous situation might be expected which exceed the present age of the universe!

Alongside the Signal Engineer's quest for safety, there has been the Railway Inspectorate (recently accommodated within the Health & Safety Executive.) For some time now, these two bodies have worked together in co-operation, and as ATP has developed in this country, a great deal of support has been given by the RI. It is now generally accepted by all concerned in the UK that ATP should meet safety criteria to the same level as that required for signalling infrastructure - interlockings, points, signals etc. Thus, vital systems shall meet probabilities of wrong (danger) side failure equal to that of interlockings. For example the generation of a permissible speed code, and its interpretation on board, used to set the intervention curve, shall be of the highest achievable integrity. However, in the UK, the presentation to the driver of the maximum permissible speed is not always considered vital. If the display erroneously advises a driver he may travel at 300 m/h instead of 30 m/h, the high-integrity ATP will apply the brakes when the train exceeds about 36 m/h. This may be worrying to the driver, but the train remains safe. I raise this point since in at least one other country - France - a quite different philosophy has evolved. Their view is that since the wheel/rail interface on modern stock cannot guarantee that no slip or slide will occur, there is no point in making the generation of the intervention curve to meet the highest integrity standards, because the measured instantaneous speed is subject to the vagaries of the wheel/rail interface. The French insist, instead that the driver's permitted speed display must be of the highest integrity, thus implying much greater faith in their drivers obeying instructions than we do in the UK. Resolution of these differing philosophies might be resolved when instantaneous speed measurement of trains no longer depends on measuring the rotational speed of axles.

Matters such as these are constantly under review by railway administrators around the world and, in this country at least, by the monitoring body, the Railway Inspectorate.

6. THE EUROPEAN TRAIN CONTROL SYSTEM

The UIC had a committee which, for a number of years, attempted to define an ATP system, basically intended for high speed lines, largely in Europe. Its work was frustrated firstly by the introduction of 3 highly disparate systems - in France, Germany and Italy - during the mid 1980's, and also by the trivial amount of time devoted to the task. The committee met for 2 days, 3 times a year. Then, the European Community agreed to help fund a manufacturer - driven programme aimed at defining such a system. This led to a re-think by UIC since the railways felt they needed to have a large part in the definition of a system of which they would be the ultimate users. A new, much more lively committee, with members providing 50% or even 100% of their time was created. It was agreed that the railways, through the new body set up within the European Railway Research Institute (ERRI) would provide functional and system specifications, and that industry, supported by the now European Union, would convert these into engineering and even manufacturing specifications.

This work has progressed to the point where Railtrack has felt itself able to grasp the nettle and has invited tenders to introduce the emergent European Train Control System (ETCS) on the West Coast Main Line.

The proposed system offers many advantages and addresses or resolves all the problems discussed above. Its key feature is that communication between ground and train will be by radio. At a stroke, this eliminates all the problems involving installation, maintenance and reliability of equipment in or about the 4-foot. It is not exactly continuous, but rather may be looked on as cyclical, since each train will have its data up-dated (by radio) approximately every 5 seconds. Although it is possible that some passive beacons will be needed, mounted in the 4-foot, so as to check location of trains, virtually all other infrastructure can be removed - mainly lineside signals and track circuits. Thus, this system can be installed with the minimum of work on the track, so reducing disruption to traffic and personal risk to staff. Once all rolling stock has been equipped with the on-board equipment, only then can existing infrastructure be recovered. However, this should be relatively painless since the new ETCS will simplify such matters as taking and giving back possessions, since the radio system will also permit track-based staff to integrate themselves with train movement instruction

7. CONCLUSIONS

ATP has had varying fortunes around the world so far, largely due to differing philosophies of railway administrations towards investment expenditure and its prioritisation. For example the UK network decided the expenditure of not a vast sum to provide door locks on InterCity coaches would save more lives than would substantially larger sums on provision of ATP statistics so far tend to prove it right.

However, as technology moves forward, and new technology such as high density radio communication between train and the control centre become available, the attraction of ATP/Cab Signalling is increasing, particularly for trunk routes.

The emergence of ETCS is providing a formal framework in which the new technologies can be integrated with the more advanced existing systems. ETCS provides advantage way beyond those of traditional ATP in areas such as staff safety, taking and giving back possessions, introduction of temporary or emergency speed restrictions, service alterations, train performance data and even commercial/pricing information. It is quite capable of becoming the centre of operating and commercial control of a complete railway.

It is now quite likely that the West Coast Main Line project can put all these features to the test. I am very optimistic that this project will lead railway safety, control and administration into the 21st century.

Figure 1 - TYPICAL BRAKING CHARACTERISTICS
(ATP Control versus NON ATP Control)

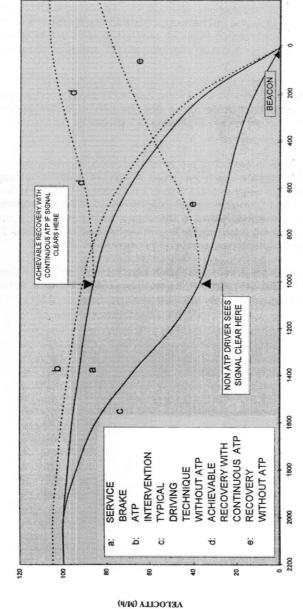

DISTANCE TO STOPPING POINT (TARGET DISTANCE) (Yards)

VELOCITY (M/H)

ACHIEVABLE RECOVERY WITH CONTINUOUS ATP IF SIGNAL CLEARS HERE

NON ATP DRIVER SEES SIGNAL CLEAR HERE

BEACON

a: SERVICE BRAKE

b: ATP INTERVENTION

c: TYPICAL DRIVING TECHNIQUE WITHOUT ATP

d: ACHIEVABLE RECOVERY WITH CONTINUOUS ATP

e: RECOVERY WITHOUT ATP

72

C511/14/116/96

Modernizing London Underground's Central Line train service

M J H CHANDLER CEng, FICE, FIEE, FIHT
London Underground Limited, London, UK

SYNOPSIS

The paper describes work that has been necessary to achieve the benefits of a large railway modernisation project once the assets have been put into service. Examples include the problems of interfacing new and old, the integration of the overall system including the "soft" issues, and the changes of culture that are required.

There are many links in the present work to the original railways which now form the Central Line. Today's passengers can still glimpse the signs of this evolution, and, in the course of providing the technology for the 21st century, today's engineers come face to face with things last seen by Victorian eyes.

1.   HISTORICAL PERSPECTIVE

The Central London Railway opened from Shepherd Bush to Bank on 30 July 1900, and to Liverpool Street 12 years later. The westbound tunnel continued beyond Shepherd's Bush in a tight 77m radius curve under Caxton Street to reach the company's depot and generating station at street level near Wood Lane. A station was opened at Wood Lane in 1908, and the track was extended in a loop around the depot to rejoin the eastbound tunnel at Shepherds Bush. This resulted in right hand running as far as the flyunder at East Acton when the line was extended to Ealing Broadway in 1920. Wood Lane was closed when White City opened in 1947. New electrified track was opened on a British Railways formation to Greenford in 1947, and West Ruislip in 1948.

Tunnels were driven eastwards from Liverpool Street before the war, but London Transport operation to Stratford, Hainault and Loughton was delayed until 1946 - 48, Epping was reached in 1949, and electrification finally replaced the steam hauled service to Ongar in 1957. The surface sections use the lines originally opened by the Eastern Counties Railway to Loughton in 1856, the Great Eastern extension to Ongar in 1865, and the Fairlop Loop from Woodford to Ilford which opened in 1903.

2.   PRESENT DAY MODERNISATION

Modernisation of the Central Line was again launched in the late 1980's with the replacement of the 1962 rolling stock and the post war signalling and traction power supply. Work is now underway to increase peak capacity to at least 33 trains per hour, with a high level of reliability, using centralised control, automatic train operation, and enhanced communication and information systems.

Examples of new technologies introduced into London Underground for the Central Line include computer controlled traction, computer based signal interlocking, and optical fibre for all communication including signalling. Systems experience being carried forward to other modernisation projects includes:

-       the need to view a metro as an integrated system of assets
-       people, staff and passengers, form part of the integrated system
-       managing major change without unacceptable service disruption
-       achieving and sustaining reliable asset performance
-       managing the interaction between new and old assets
-       acquiring new skills and cost effective specialist product support.

In 1987 London Underground regrouped railway operation into line based business units. Stewardship of most assets, except the main power supply, was devolved to the lines from 1992 in order to increase accountability for service delivery. This has provided a single point of focus for system integration and improvement of asset performance, removing the classical division between engineering disciplines.

3.   ASSET PERFORMANCE

It was soon clear that new technology can worsen reliability, for example:

-       Newly developed products near the top of the "bathtub" curve
-       Insufficient ruggedness of new components in critical functions
-       Additional complexity of "nice to have" features
-       Inadequate specification of total system performance.

The end of the story has not therefore been the delivery of the individual assets against project management milestones. User driven activity has been necessary to ensure that the overall functionality is achieved with the required level of reliability once the assets are embedded in the operational system. This is likely to continue until around 2000.

Fig 1 shows how top level service expectations are cascaded to asset performance measurement and improvement plans in a closed loop process.

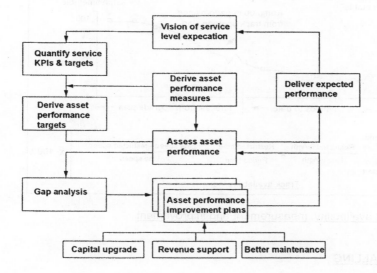

Fig 1: Asset performance improvement process delivery

Individual project managers have been assigned to the major asset improvement plans, and each plan includes a forecast of improved performance with credibility levels.

The number of faults, in conjunction with delay, has been the traditional measure of engineering performance, but delay can include the operational element of service recovery. Asset availability has therefore been used as a truer measure of both equipment and maintenance performance since it combines fault rate, response and repair time. The calculation has initially been kept simple and free of weighting factors in order to get reliable collection and processing of data underway, and establish ownership by the individual asset managers.

Fig 2 shows an example of the availability definition for track. Asset performance is managed using the left hand scale for the current timetable, which shows a current average of around 98% due to temporary speed restrictions. The right hand scale shows the additional demands of the 33tph timetable against which the forecast improvement from the asset upgrade plan is plotted.

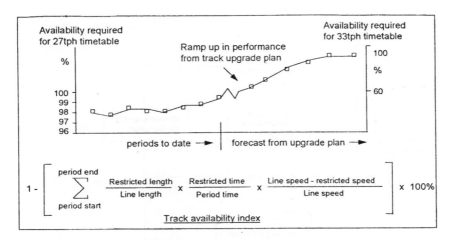

Fig 2: Track availability: measurement and improvement

4. SIGNALLING

The old signalling on the Central Line dates from the 1940s or earlier, and is operated conventionally by lever frames and by passing train descriptions between boxes. Trains are protected by pneumatically operated train stops and train mounted tripcocks.

Trains will be driven automatically (ATO) in the new system, and under normal conditions the central computer will schedule all main line movements. Automatic Train Protection (ATP) codes, station to station run profiles, correct side door enabling information and platform CCTV signals are picked up by the train from trackside equipment. The train can also download fault data.

The modernisation project plan has introduced the new trains and signalling over several years. This provided limited-risk steps, but has involved the line in numerous interim stages:

- inter-running of new and old trains
- new relay based signalling
- processor based interlocking at the later sites
- inter-running of tripcock and code based protection
- local control, centralised control and automatic train operation

The project plan did not contain signal maintenance provisions. The most appropriate solution post commissioning has been to retrain line based staff to deal with the trackside systems, including the processor based interlockings, and to seek bought in support for the centralised control and communications systems and software.

Existing staff have been keen to become licensed in the new technology. A professional trainer was added to the team who developed a leading knowledge of the technology and close liaison with the signalling contractor. Remote diagnostics are due to be provided to the control centre.

The asset upgrade plan for signals is derived from the most frequently occurring faults, and addresses the quality of relay manufacture, the durability of track connections and cable protection, the design of power feeding arrangements. Processor based interlocking has provided good reliability of the vital logic, most of the difficulties have been associated with the development and reliability growth of the external interfaces.

Fixed block signalling has been retained and the jointless track circuits carry the ATP codes. Intermittently poor electrical properties of the old track have had to be painstakingly eliminated, particularly at points and crossings.

Signal availability dropped to 96% due to the high fault rate, and the aspirational target was dropped from 99% to 97%. Shorter repair times have achieved an improvement to 97% and the target for the current upgrade plan is 98%.

5. TRACK TO TRAIN INTERFACES

The Central London Railway tunnels driven in 1896-98 had an internal diameter of 11ft. $8^1/_4$in. with bridge section rails fang bolted to longitudinal timbers, and a third, central conductor rail. 10,000 rings were enlarged in the 1935-40 new works programme, when a fourth positive rail was added, albeit at a non standard height due to the still limited clearance.

This inheritance plagued the introduction of 1992 tube stock (92TS), when, again, a complete kinematic survey was required, and further adjustment of tunnel rings and removal of other gauge fouls proved necessary. Tight tolerances on relevelling the active suspension are necessary following lifting of the cars, and this is a critical factor in achieving the 95% train availability needed for the higher service level.

92TS shoes are mounted on two pivoted frangible links and do not rotate about an axis parallel to the track as on earlier stocks. The breaking strength of the links, and the tolerance of shoe width to conductor rail position have proved critical, and have had to be adjusted in practice. Train borne video surveys have provided a valuable understanding of how track features affect the vertical acceleration and horizontal position of the shoes.

The trains were designed to prevailing track standards, and were factory tested on rigs fed with actual track profiles. However the interface between old and new has to be finally resolved in operational service with the possibility of serious shoegear-to-rail incidents. The whole of the Central Line conductor rail has been inspected and fettled to reduce such risks.

The changed interface between wheels and running rails has also had to be managed whilst minimising the impact on service. Improved track lubrication was necessary to cope with mixed wheel profiles during the interrunning of the new and old fleets, and wheel flange lubrication has subsequently been introduced.

Track is being upgraded to enable the line speed to be increased for the new ATO timetables, equating to a phased increase in track availability of 25% (see Fig 2) Another historical inheritance, the very tight curves at Shepherds Bush and Bank, have prompted a review of gauge, and the use of harder grade rail, to combat sidewear. Given consistency of speed in ATO, tracks are now recanted to equilibrium where clearances permit, rather than to the previous standard with deficiency.

A test train has been run at the new line speeds equipped to measure vertical and horizontal acceleration of the wheelset, bogie, and car in order to provide calibration for the new vehicles with portable in-car recorders, and the LU track recording vehicle data. The data is also linked to computer referenced underframe video recordings. This has identified those track defects which give rise to insufficiently damped lateral resonance, which, aside from ride comfort considerations, would increase train maintenance and impact availability.

6.   TRACK TO FORMATION INTERFACES

Poorly graded and unstable embankments, usually of ash, are another historical inheritance of time and cost constraints during the expansion of Victorian suburban railways. The Central Line has a high proportion of such structures, with the consequential risk to the reliable achievement of increased service levels. The upgrade of assets as an integrated system has therefore included a programme to remedy earth structures with the poorest factors of safety, starting at around 0.75.

Structural solutions have been generally been used where the tipped material is poorest, earlier stabilisation methods having failed. In the worst cases, a closely spaced line of small diameter reinforced concrete piles is formed along both shoulders of the embankment, bored to below original ground level, each carrying a reinforced capping beam. These are tied across the embankment by rods grouted into tubes below the track bed. The structure is self supporting and addresses loss of shoulder and rotational slip. Mini piling rigs operate without possession from a fenced off platform, and only the upper level of vegetation needs to be cleared.

Gabion walls, a toe drain and regraded slopes have been used where boundaries are wide and it has been possible to remove vegetation. In simple cases counterforted drains have been sufficient.

7.   PEOPLE AND INFORMATION

People behaviour and adequacy of information are essential elements of system integration. For example, passengers and staff form part of the dwell time process which the automatic system seeks to control second by second. Training of station staff as well as train staff has been vital.

Instinctive manual intervention at the control centre, if permitted, may well spoil the overall optimisation of automatic train regulation. Overcoming this culture has been made difficult by having to retain various degrees of manual control during the changeover process.

Greater precision of train service delivery, and ongoing system development, demands performance data which reveals the exact operation of the system in real time such as time / space diagrams of successive trains, and distribution plots. This, in turn, requires more statistical awareness and imagination in creating new measures, and close integration of the line's operational and system engineering staff.

Other information systems include on-train diagnostics, maintenance management updated automatically from train data, and cross linked documentation storage for all systems accessed by a Yourdon model. Getting the systems to add real value to the process each is intended to assist, and inserting and updating the data, have proved most difficult and prone to over ambitious specification.

8.   CONCLUSION

The challenge of such projects extends beyond the delivery phase into making them work really well in the operating environment. The initial excitement, and large scale spend is largely over is over at this stage, and project staff can be distracted by new opportunities. Getting the detail finally right is critical to success, and it is important not to let standards slip when the inevitable difficulties occur.

Reference

Follenfant, H G, Reconstructing London's Underground, 1974, London Transport.

C511/14/124/96

Material management and after sales support

A E ORPE MIMgt
ABB Daimler-Benz Transportation, Derby, UK

The supply chain for spares in the rail industry is cumbersome and costly with responsibility borne largely by the operator to ensure continued availability of his vehicles. It is now essential for suppliers to demonstrate their commitment to support their products in service and not only for the warranty period. This paper examines a number of ways in which a supplier can assist the operator by adding value to the process whilst giving value for money.

1.   INTRODUCTION

With the evolving rail industry seeking to examine the cost drivers within each business the cost of holding stock to support its vehicles is recognised as an area which can be greatly improved by the suppliers taking a more pro-active role.

There is a wide spectrum of services which suppliers can offer ranging from satisfying ad-hoc requirements through to taking overall responsibility for the complete material support as part of a full vehicle maintenance package. The supply and subsequent management of materials is a key element in the successful operation of a rail service but all too often is largely ignored when identifying means of improving total business performance.

It would be a mistake to concentrate on the complexity of the subject and lose sight of what it is we are trying to achieve. We need to use the building block approach to our thinking with the simplest solution being our objective. By giving clarity to the solution we are more likely to gain acceptance at all levels of the organisation in which the problem exists.

2.   THE TRADITIONAL METHOD

Before we consider some alternative methods of supply it is important we recognise how vehicle manufacturers were traditionally required to support the introduction of a new vehicle into service with spares that would ensure its availability to meet the operators requirements.

As part of the new vehicle order all main spares would be identified and ordered in quantities deemed sufficient to meet future demand for insurance and overhaul purposes. Delivery would be phased with a small quantity available with the first vehicle and completion to co-incide with the last vehicle. This early purchase was considered viable as the benefits of buying with build ensured the best price and instant availability of what would be a long lead item.   One of the potential problems caused by the early supply of spares was available storage capacity which was clearly demonstrated some years ago when spares required for a vehicle in operation south of the Thames were stored in Scotland.

In addition to the main spares the contractor was also required to produce a Recommended Spares List in sufficient time for the customer to order and the spares to be available once again with the first vehicle.

With most vehicles the compilation of the list necessitated starting from scratch as there was usually very little component commonality from one vehicle to another. To give an idea of the scale of this exercise let us consider a typical Diesel Multiple Unit. The number of components required to manufacture the vehicle would be approximately 10000, of which some 1600 would be considered as either consumable or insurance spares to be stocked by the customer. Of these, only 80-100 could be classed as regular user items which leaves a balance of 1500 items with a small or zero turnover but many would be potential vehicle stoppers if they were not readily available.

The criteria used by the customer to decide the quantity to be ordered were many and varied. The manufacturer would recommend a quantity based, for example, on predicted failure rate, repair turnround time and replenishment lead time. The customer may have ordered the recommended quantity or been driven by other factors such as budgetary limitations or in some instances by available storage space.

The one common theme with this traditional method was that the customer took all of the risk and responsibility for financing, ordering the correct parts and stocking. Even worse, the customer was forced to accept that only he would stock the spares with the manufacturer holding little if any support stock with the resultant production lead times effectively financed by the customer.

Obviously this situation could not and has not continued and todays operators are rightly expecting far more support and risk sharing by the suppliers.

The one constant we are all currently experiencing in the Rail Industry is change and it is essential that we harness the energy which change can bring about and use it to our mutual advantage.

I mentioned earlier the need to introduce a simple systematic approach to the problems and the following are some examples where Adtranz have concentrated on listening to the customers concerns, devising solutions and at the same time adding value to the whole material support process.

3. INJECTORS

As part of a Spares Contract we supply injectors for replacement at the depot. When the injectors are changed there are 3 washers that are necessary to carry out the job. All of these items were stocked separately which involved 4 stores locations, individual stock control and re-order processes and time for the storeman to pick and give to the fitter. The simple solution was to supply the injector and washers shrink wrapped to a small card and supplied against one part number. This small example gave us the start point for using the same principle on far larger applications.

4. MAINTENANCE KITS FOR B EXAMS

A great deal of time and effort is spent on ensuring all of the spares required to carry out a specified B Exam are available when the work is programmed to take place. Working with the operator we agreed the material content for each examination from B1 to B12 then developed specialised packing so that the consumable material for one exam is all contained in one box. The on-condition material was treated differently as it would not be sensible to transport unused material around the country. The depot holds stocks of the on-condition material, the level being determined by assuming 100% change out on the heaviest exam week. The stock is checked weekly and all consumed items are automatically replenished using a two bin system.

The total service was agreed and competitively tendered as value for money remained paramount to the operator.

The service and accrued benefits can be summarised as below.

THE SERVICE

- Scheduled, once a week, supply of kits to meet agreed Maintenance Programme.

- Day and time of delivery to suit the customer.

- Two Bin System for on-condition items.

- 100% Service Level.

- Rapid ad-hoc service.

- Spares Support Engineer on a regular or as required basis.

- Dedicated contact personnel.

- Technical Support - Product Improvements.

- Flexible response to meet customers changing demand.

- Regular Review meetings.

- Partnership approach.

THE BENEFITS

- Prices fixed for 12 month periods - budgetary certainty.

- One part number per kit.

- Reduced stock holding - increased stock turnover.

- Reduced manual handling.

- Reduced storage space.

- Prices are all inclusive.

- No surplus and obsolete considerations.

- Suitable enclosed box per kit, palletised to enable ease of movement to vehicle side.

- Reduced material losses.

- Supplier has detailed vehicle knowledge.

- Own transport and drivers used - understand the product and the service.

- Penalty for non-performance.

- 12 month warranty.

The contract has now been in operation for 12 months with a 100% service level maintained throughout the period. A number of other operators are interested in introducing this user friendly concept and are considering proposals.

5.   SPARES SUPPLY CONTRACTS

Another area which has been the subject of some development has been the contractual supply of nominated spares against agreed performance and service improvement criteria. The underlying theme of such contracts should be a partnership approach which involves both parties working closely together to deliver tangible benefits for all concerned.

Typically the operator would consider the following elements to be an integral part of the contract:-

- Fixed prices.

- Regular Performance Review meetings.

- Penalties for non-performance.

- Termination for non-performance.

- Technical Support.

- Emergency access to materials.

- Support stocks held by the supplier.

- Rapid response to enquiries.

A recent contract calls for the supplier and the customer to work together to maintain continual improvements to supply with both reduction in use and cost during the period of the contract.

The following areas of improvement indicate how far customer thinking has developed from the 'buy cheapest' approach which prevailed for far too long in the industry.

- Reduced lead times.
- Improve delivery performance.
- Quality - zero defects.
- Inventory Levels.
- Administration Costs.
- Product Development and Modifications.
- Unit Costs.
- Material Handling (move towards kits).
- Standardisation across classes.

Why not develop the theme further and let the supplier finance the customers stockholding with invoices raised on an as used basis. This would focus the suppliers efforts into concentrating on the shortest route in the supply chain with the shortest time taken and subsequently less money involved in financing the process.

6.   PRODUCT IMPROVEMENTS

As operators become increasingly isolated and focused on meeting the needs of their own customers it may result stagnation of their vehicles as their energy is expended on making what they have work properly. There is an increasing role for suppliers of a range of vehicles to read an improvement in one vehicle across to a different vehicle with a different operator.

The supplier would benefit with increased sales and the operator by enhancement to either reliability, safety, material consumption or cost.

Whilst this should be obvious to all parties I do not believe it is something that will automatically happen.

7.   REPAIR OF COMPONENTS

None of us would tolerate the alternator on our car failing and being told by the main dealer to come back in a week while it is repaired and incidentally no replacement vehicle is available. Yet this is precisely what happens regularly with rail vehicles. The repairers have to bend their minds around this problem and regard it as an opportunity and not a threat. They should offer a service exchange facility to enable unit exchange to be carried out locally thus keeping vehicle down time to a minimum.

8.   CONCLUSION

As with any betterment there is sometimes a price to pay. The key consideration is total price against total cost and as the pressure increases on providing more reliable and frequent services, then the cost of not having the resource available is often far greater than the cost of providing it.

We do not have all of the answers but by adopting a customer focus strategy we are able to concentrate on the areas of biggest impact with value for money being the ultimate goal.

C511/17/068/96

Factors influencing derailment risk

C A O'DONNELL and **A K CARTER** MIMechE
B R Research, Derby, UK

1.   INTRODUCTION

When a train derails the result is usually extremely costly in terms of damage to vehicles, track and infrastructure. In addition there is the resulting disruption to services, the potential hazard to surrounding areas and the potential contamination of the environment, not to mention injury/loss of life to staff, customers and third parties.

This paper will examine the effect that vehicles and track each have, as they interact as a system, and how their conditions influence derailment risk. Such conditions are governed by track and vehicle parameters and hence owners, maintainers and operators in each sphere need to be aware of those factors, which are within their control, that can have a significant influence on derailment risk. Such an approach is essential if the whole range of companies involved in the various aspects of railway operations are to collaboratively reduce the number of derailments.

Ideally all parties involved should have a comprehensive understanding of how the safety requirements are to be met. These safety requirements are well defined, for railway systems within the UK, by the various documents issued by Group Standards.

In particular such parties need to develop an awareness of those issues within their sphere of control, such as maintenance limits that need to be met. Further they need to understand the mechanism of how such limits are regulated and their impact and influence on other areas. A knowledge of how individual maintenance factors effect derailment proneness is therefore needed. Such a combined approach will allow the individual groups effectively to target strategic aspects, and within increasingly stringent financial restraints ensure continued safety against derailment risk in the most cost effective manner.

In order to apply the foregoing it is vital to appreciate those derailment mechanisms over which some degree of control rests with either the vehicle or track. Further it is important to understand exactly what part the various elements play in each type of derailment mechanism. Only in this way will it be possible to know what aspects to consider when deciding the degree of risk that applies in the widely varying situations met in practice and thus be able to take proactive action.

2.   DERAILMENT MECHANISMS

Derailments occur due to a multitude of causes but those we will be considering in this paper will be limited to situations where faults in the track and/or vehicle are a primary mechanism of derailment. Those mechanisms we shall examine in detail are:–

Flange climb.
Track failure.
Vehicle response to track features.
Geometric/dimensional mismatch.

3.   FLANGE CLIMB

3.1   Mechanism of flange climb derailment

Flange climbing as a derailment mechanism has been recognised for a long time. It is now nearly ninety years since Nadal postulated his formula expressing the limiting ratio of the lateral force (Y) over the vertical force (Q) as being governed by two factors, namely the angle (α) at the point of contact between the wheel flange and rail head and the coefficient of friction (μ) at this interface. i.e.

$$\frac{Y}{Q} = \frac{\tan \alpha - \mu}{1 + \mu (\tan \alpha)}$$

Although this expression is a pessimistic approximation (for example it takes no account of the wheelset's angle of attack) it is still a useful guide. The main consideration that it illustrates is that, for a given flange angle and coefficient of friction:–

INCREASE IN Y/Q RATIO ■ INCREASED RISK OF DERAILMENT

This shows that either a reduction in the vertical force (Q) for the same lateral force (Y) or an increase in the lateral force (Y) for the same vertical force (Q) increases the risk.

Derailments where there is a significant reduction in the vertical force (Q) at a particular wheel are the more common and are frequently associated with load transfer from one side of the vehicle to the other.

Such a transfer of vertical force can ensue from poor vehicle set–up, incorrectly adjusted or faulty suspensions or uneven loading of the vehicle. Similarly track twist will result in wheel off loading. Track twist exists at the entry to and exit from curves due to the design transition from flat tangent track to canted curved track. It also occurs due to faults in the track such as local dips, which occur as a result of voids in the ballast.

Derailments attributed mainly to an increase in lateral load (Y) are generally less common. However, the lateral force will increase if vehicle wheelsets are misaligned in a curve, which can result from differential wheel wear. Lateral forces are also increased for vehicles which are dynamically unstable. Similarly curves with alignment faults can significantly influence the lateral force.

When considering the limiting Y/Q ratio it will be understood that the coefficient of friction (μ) cannot be totally controlled and hence values of friction normally encountered in operational service have to be allowed for i.e. generally within the range from $\mu = 0.1$ to $\mu = 0.45$. Flange angle on the other hand is defined when profiles are turned and will tend to wear such as to produce a steeper angle and a reduced risk and therefore this element can be controlled.

On BR, in common with the UIC requirements, a value of Y/Q of 1.2 is considered the limit.

Flange climbing can also occur where there are discontinuities in the rail shape, for example the blending in of side worn to new rails can present an approaching wheel with an effective ramp that will assist it to flange climb. This mechanism is different from that controlled by limiting Y/Q.

3.2   Influential factors in flange climb derailments

The influencing factors on Y and Q, and therefore on derailment, can be summarised under two headings:–

3.2.1   Reduced vertical force

Cant excess
Designed track twist (eg curve transitions)
Track twist faults (eg dipped joints)
Poor track support (voids)
Poorly designed vehicles (high twist stiffness)
Badly set–up vehicles (eg suspension clearances)
Twisted vehicle frames
Badly loaded vehicles (side–to–side or end–to–end)

3.2.2   Increased lateral force

High angle of attack (increases with reduced curvature)
High coefficient of friction
Sudden change of lateral alignment
Blending in of side worn to new rails
Stiff yaw suspension
Poor rotational performance of bogies (about the bogie pivot axis)
Increasing wheelbase
Poorly aligned wheelsets
Uneven wheel diameters

4.   TRACK FAILURE

There are two main modes of track failure namely gauge spread and track buckle.

4.1   Mechanism of gauge spread derailment

Gauge spread occurs where the restraint of the rails is unable to withstand the forces applied by the wheels. Some vehicles exert higher forces than others and the force levels also change with levels of wheel–rail friction. However it must be appreciated that some level of gauge spreading force is inevitable (particularly on sharp curves) and therefore the track structure must be able to withstand this.

When the method of securement of the running rails reaches a point where it can no longer withstand the system of lateral forces pushing the rails apart, the result is that the rails are pushed and/or rotate outwards beyond the point where the wheelset can be supported. This allows one or both wheels to drop into the four–foot. Such a derailment will occur only where the dynamic track gauge (i.e. under load) increases by some 85mm from nominal.

It is important to bear in mind that the potential length of track where initial failure occurs can be a distance of only three sleepers (1.5m). One of the obvious factors when considering the risk of gauge spread derailments is the static gauge. Probably less obvious (and not readily measurable) is the dynamic gauge i.e. the gauge under the dynamic load as vehicles pass. This will tend to be worse for vehicles having a long wheelbase or multiple axle bogies (i.e. three or more axles).

Potential sites at which gauge spreading derailments could occur will normally be known, having been identified from certain maintenance measures already undertaken. Such sites will usually exist on curved track (the highest risk on any particular curve tending to be at the point of tightest local curvature). There will be evidence of loose or broken rail fastenings, split timber sleepers, movement and fretting of sleepers under and to the outside of baseplates and chairs. There will usually have been attempts at temporary strengthening via the use of tie bars. Over a period of time a wear pattern emanating from wheel tread corners may occur on the rail head.

On bridge sections the support to the rails provided by way beams is another potential risk that can easily be overlooked. Similarly at such sites it is not usually possible to tamp right through. The effect of tamping up to a bridge section from either side can result in somewhat deficient alignment which will aggravate any potential problem by resulting in tighter local curvatures and hence increased lateral loading.

Increases in vehicle yaw suspension stiffness and in the axleload will aggravate the gauge spreading loads imposed upon the track. The lateral gauge spreading load imposed will be proportional to the coefficient of friction and increase as μ increases, and also as the curvature reduces. Finally any reduction in the integrity of the track fastenings or lateral track bed support will serve only to worsen the situation.

As speed, and hence cant deficiency, increases the gauge spreading force does not generally change significantly. The reason for this is that as the speed over a given track section increases, there will be a corresponding weight transfer of load to the outer (high) rail but, conversely, the load on the inner (low) rail reduces. The product of these vertical loads and the coefficient of friction at the rail head is a major factor affecting the lateral force but an increase on one rail is accompanied by a reduction on the other. Hence although the track shifting force increases significantly with speed, this is not the case with the gauge spreading force.

For similar reasons, merely changing the cant of the track at a particular curve does not significantly affect the gauge spreading force, as the system of forces generated by the vehicle is effectively contained within the wheel–axle–wheel–rail–sleeper–rail system.

4.2 Influential factors in gauge spread derailment

The following factors are significant in the avoidance of gauge spread derailments:

Dynamic gauge
Integrity of track fastenings
Track bed/support
Curve radius
Vehicle/bogie wheelbase
Axleload
Yaw suspension stiffness
Coefficient of friction on the rail head (although this is not controllable)

4.3 Mechanism of track buckle derailment

This type of derailment occurs due to excessive longitudinal compressive forces built up in the rails as a result of their expansion due to high temperatures. The risk of a track buckle developing at any location increases with reduction in stress–free temperature, higher rail temperature and a reduction in any expansion gaps.

By disturbing the track bed the potential for buckling will increase. Such a situation can also be brought about by insufficient lateral support (resulting from lack of ballast) and changes in lateral track stiffness resulting in local "weak" points. Such points can occur between areas of pointwork, bridges, etc.

With track buckles it is important to note that the vehicle only acts as a "trigger". This is provided by a combination of the track being lifted by the "precession wave" and the lateral force generated as the vehicle passes a potential site. At the extreme limit and given the worse possible scenario, track will buckle even without such a "trigger".

4.4 Influential factors in track buckle derailment

The following factors are significant in relation to track buckle derailments:
Correct stress–free temperature
Maintenance of expansion gaps
Disturbed track bed
Insufficient ballast
Changes in lateral track stiffness

High rail temperature is the major factor but it is not controllable

5. VEHICLE RESPONSE TO TRACK FEATURES.

The main mechanism of derailment attributed to track features is that due to cyclic top which is the generic title given to long wavelength faults in vertical track top.

5.1 Mechanism of cyclic top derailment

Historically this has been understood to be a fairly continuous series of even dips extending over several cycles. Recently however there have been derailments attributed to a similar mechanism but in these instances the fault extends over only two to three dips which are seen to increase in severity. Such a fault has colloquially been referred to as a "ski-jump". The long wavelength faults in vertical top are normally associated with wet spots in the track bed.

The risk of a derailment caused by cyclic top increases as the input frequency, from the cyclic dips, approaches the vehicle's natural vertical response frequency in pitch and/or bounce. This situation usually occurs for a two axle vehicle with a wheelbase approaching some sub-multiple of a standard rail length and with a high frequency response i.e. a result of a high vertical suspension stiffness to low mass ratio. Where the vehicle's natural response frequencies are higher, the co-incident speed will also be higher and the risk of derailment will be greater due to the fact that more energy will be contained within the system.

Similarly, the vertical damping rate is important as this assists in controlling the rate of energy dissipation and with low levels of damping the situation will worsen.

The response to the vertical cyclic dips by themselves will not be sufficient to cause derailment as the wheelset will also need to be forced laterally at the same time. However, as already discussed such lateral forces may well be easily generated by curving or even by the vehicle's own natural lateral response to normal track inputs.

5.2 Influential factors in cyclic top derailment

The following factors are significant in relation to cyclic top derailments:

Long wavelength faults in rail top
Wet spots
Vehicle wheelbase
Dynamic vehicle response
Operating speed

6. GEOMETRIC/DIMENSIONAL MISMATCH.

If a geometric or dimensional mismatch occurs between the wheel and rail it can lead to derailment. In particular this is the situation at pointwork, where either flange climbing or splitting of the switch will cause derailment.

6.1 Mechanism of flange climbing derailment at switches

The risk of flange climbing derailment at a facing switch can be aggravated if the stock rail is heavily sidecut (as can be the situation if the switch is usually used in the through direction and also lies on a sharp curve). This effectively prevents the switch blade seating correctly and will present the approaching wheel with the switch blade edge.

This situation worsens yet further if the switch blade is incorrectly profiled (in cross–section) as the edge presented may be sharp and hence "dig" into the wheel flange. This is particularly the case with switch blades manufactured from manganese and mill heat treated steels which exhibit a natural tendency to wear to a sharp edge.

6.2 Influential factors in flange climbing derailment at switches

For flange climb derailment at switches the factors are similar to those for plain track flange climbing (Section 3.2) but with the additional effects:-

Side cut stock rail
Switch blade profile
Switch blade material

6.3   Mechanism of switch splitting derailment

Vehicles will derail if a switch is split by the action of a wheel that separates the switch blade from the stock rail. This then would allow the wheel to pass the wrong side of the blade and run along the stock rail until the now diverging rails allow the wheelset to drop in.

For a wheel safely to negotiate a facing switch the geometry needs to be within defined limits. If not, there is the distinct possibility that a wheel or wheels will take the "wrong" route, and hence split the points. Derailments resulting from the splitting of points are likely to occur where a gap exists between the switch blade and stock rail. This may be due to faulty locking, slack or play in the mechanism, an insecure stock rail, incorrect components, debris contamination between switch blade and stock rail or if the wheel flange has faults i.e. thin flanges and/or toe radius build–up.

6.4   Influential factors in switch splitting derailment

The following factors are significant in the avoidance of switch splitting derailments:

Gap between switch blade and stock rail
Faulty locking of switch blade
Slack or play in the mechanism
Insecure stock rail
Incorrect components
Debris contamination
High angle of attack
Wheel flange faults

7.   DISCUSSION

It is important to put any discussion of derailment risk into an appropriate context. The railway is a very safe system of transport, and this paper is not intended to suggest otherwise, but we cannot afford to be complacent.

The yearly derailments statistics for the last ten years show that, within certain limits of annual fluctuation, the average number of incidents has remained fairly constant. Considering only those derailments whose mechanisms are covered by this paper it is evident that the overall trend for derailments due to vehicle faults is steadily falling. For derailments where track was at fault the overall trend has remained roughly level.

It is evident from the statistics that there is still room for improvement in all areas and there are lessons to be learnt, and acted upon. It is hoped that this paper will have contributed to that process by laying out and highlighting some areas and factors that can form the basis for such action.

C511/20/126/96

Refurbishment of tube trains

M D THOMAS MSc, FIMechE, MIEE
Bombardier Prorail, Wakefield, UK
D MORPHEW BSc, MIMechE
London Underground Limited, London, UK

SYNOPSIS

Bombardier Prorail are now well into the refurbishment of London Underground's Piccadilly Fleet of Trains. This refurbishment is part of the ongoing upgrade of London Underground System and is the most extensive to date. This paper summarises the background to the refurbishment programme, the selection process of the original contractor, the work content and the solution developed.

A key feature of this work programme has been the partnership between Piccadilly Line Project Team and Bombardier since the transfer of the work from RFS Doncaster to Bombardier during 1994.

REFURBISHMENT OF UNDERGROUND TRAINS

1.0 BACKGROUND

Since 1989 London Underground has undertaken a fleetwide programme of improvements to passenger rolling stock in order to conform with the LUL Code of Practice for Fire Safety. This requirement primarily concerns the replacement of identified hazardous materials, however work completed to date shows that by extending the scope to include interior refurbishment, significant project synergy, and commercial, financial, and passenger benefits can be achieved.

An assessment of the compliance of rolling stock prioritised the urgency of the replacement works and this has resulted in a stock by stock refurbishment plan. The 1959 and 1962 tube stock fleets, operating on the Northern and Central Lines, have undergone modifications to replace non-compliant materials only. The first full fleet refurbishment carried out was ambitious, comprising the 603 cars of the 1967 and 1972 tube stocks; these are very similar vehicles and, apart from line identity on the interior liveries, were refurbished in a similar manner by Babcock Rail between June 1990 and March 1995. The 276 'C' stock cars were refurbished over a similar period by RFS Engineering of Doncaster; the scope of this project was wider than that of the 67/72ts as a major interior reconfiguration was carried out and for the first time included what was going to become a very popular feature, the provision of car end windows to improve passenger security. 'A' stock, built in the early 1960's for the Metropolitan Line is currently receiving similar refurbishment by Adtranz in Derby.

One major feature of all refurbished stock is the new corporate red, white, and blue livery; previously in unpainted aluminium the stock required periodic acid washing and was very susceptible to graffiti which proved impossible to remove completely due to the porous surface.

The 1973 tube stock was built between 1974 and 1976 by Metro-Cammell Ltd and provides all the service trains for the Piccadilly Line. It comprises of 522 cars formed into three car units, two units making one train. The Piccadilly Line has 67 route km. and carries some 494,000 passengers per weekday; the 87 trains run a total of 10.5M train km per year, some 18% of the total network train km. As well as serving the central areas of London it provides a vital link to Heathrow airport. The requirements of the Heathrow service again put a different set of requirements into the refurbishment scope; passengers needed to be able to store luggage without compromising door space.

In order to gauge passenger reactions to some of the radical ideas being put forward a three car unit was refurbished by Metro Cammell in 1990 to an interior design by Jones Garrard. Each car differed in interior detail, featuring a number of innovative features including luggage racks, smaller cross seats, various configurations of 'stand back' areas, and several combinations of grab poles and rails. A formal market survey was carried out at three locations, each representing a different customer base; passengers were invited to sample each of the three interiors and then complete a thorough questionnaire covering both physical layout and ambience. There was inevitably some conflict; passengers from the Rayners Lane branch were not enthusiastic about the loss of seating to luggage stacks, whilst Heathrow users rated them highly! The end windows again proved popular, but not so the smaller cross seats soon christened 'mother and child' seats. The questionnaires highlighted demands for better information, clean trains, and good lighting.

A workshop was held to establish the definitive scope of work, this involved maintainers, engineers, user representatives and of course referenced the market research data. Features included in the scope were:

- all cross seats replaced by longitudinal
- large 'multi-functional' standbacks by double doorways
- a number of 'perch seats', to combat the reduction in conventional seating
- passenger emergency alarms with 'talk-back' facility.
- interior dot-matrix passenger information displays.
- vastly improved passenger audio information.
- full forced air saloon ventilation.
- enhanced drive environment including cab air cooling
- fitting of impact resistant windscreens
- fitting of underframe 'arc-barriers'
- increased standby battery capacity
- improved 'detrainment' facilities

Warwick Design were chosen to carry out the interior design of the vehicle and produced a full size half car interior mock-up. The design developed that produced for the new Jubilee Line stock and established a theme that would eventually continue into the new Northern Line stock currently in design. The mock-up eventually formed part of the invitation to tender representing LUL's aesthetic aspirations and it rested with the selected contractor to translate these into an engineered solution.

Once the scope had been finalised the technical specification was completed and combined with the terms and conditions and issued as an invitation to tender to six potential contractors that had already passed the pre qualification stage. Prior to the return of the tenders and, in order to ensure objectivity in the evaluation, a weighted scoring system was devised. One bidder dropped out and at the end of ten weeks five generally compliant bids were returned for evaluation. 'Menu' pricing allowed accurate comparison of costs to be made and where particularly high or low prices for a work package were identified, when compared with the other bidders, the particular contractor was requested to explain the reason for the cost anomaly.

Three tenderers went through to the second stage. During the first round over 200 technical and commercial questions were raised by LUL, it was therefore decided to re-invite the tender against a revised specification which included the clarifications received during the first round. Commercial qualifications were discussed with the tenderers who were advised to submit unqualified bids. The third round required the bidders to submit revised breakdown sheets for each area of the specification to reflect technical and commercial clarifications of the second round bid. They were also required to complete annual payment phasing details so comparisons could be made with LUL's cash availability. On 5th May 1993 the contract, worth approx. £75M, was awarded to RFS Engineering of Doncaster and the first train arrived at the works on the same day. controlled stripping out and design work started almost immediately. Regretfully, RFS went into Administrative Receivership 15 December 1993. Following extensive negotiations with the Receivers LUL agreed not to terminate the contract but continue to trade via the receivers with the intention of transferring the contract to a new owner. This was achieved in April 1994 when the RFS passenger business was sold to Bombardier Prorail; the LUL contract being transferred at the same time.

2.0   PROJECT OBJECTIVES

The objectives of the refurbishment were:

1. Enhance Passenger Environment
2. Replace materials used in the original construction which were non-compliant to the current Fire & Safety Regulations.
3. To improve the CAB environment for operating staff and provide information and data systems for Passengers, Operators & Maintenance Staff.

In addition to the primary objectives an overiding secondary objective was that the car weights should not be increased due to the refurbishment.

3.0   MODIFICATIONS & REPAIR OF BODYSHELL

The main roof modification involved cutting apertures to house the seven saloon ventilation fans and 14 outlet ducts.

On the inside of the car mounting point bracket for the old 'ball type' passenger handholds are removed and replaced by a similar purpose designed casting to receive the grabrails and poles of the new design.

At the car body-ends, window apertures are cut to a quadrant 'D' design reflecting the car body profile.

In the central bays, formally occupied by companion seating, a new centre seat bay frame provides for four along car seats to each side. This framework acting as the structural link for the bottom of some of the grabpoles.

These centre bays are shorter between the draughtscreens to provide deeper multi-purpose standbacks. The floor sub-plates are renewed here to optimise the floor level profile.

The floor and floor traps are replaced by 'aircraft style' aluminium honeycomb construction. This in turn covered by 4mm thick Vamac matting, features within the floor design provide the bottom anchorage for the central grabpoles, the top anchors in these cases being an inner roof place 'spreader'.

Because of the increased tare weight resulting from equipment such as the static converter (cars are approximately 1.5 tonnes heavier in tare) an FEA was commissioned to look at the main underbody frame. On the trailer cars the static converter is mounted longitudinally 'amidships' but offset from the longitudinal centre line, housed between the sole bar and the car main frame.

4.0   INTERIOR DESIGN

4.1   General

The first 1973 Piccadilly Line Train entered service on 19th July 1975, and the complete fleet has a design life of 40 years, up to 31st December 2015. The replacement interior has to be designed to reflect the remaining 19 -> 20 years of life of the fleet as well as meeting the requirements of both the LUL code of Practice for Fire Safety of Materials and the LUL resistance to graffiti specification.

4.2   Interior

The train interior has been stripped out and totally replaced to improve aesthetics, update the Line image, increase floor area, and incorporate new technologies. The original transverse seating in the centre bay has been replaced by longitudinal seats. Perch seats have been installed at the standbacks and bodyends. Bodyend windows have been provided. With the change of seating layout in the centre bay, the adjacent standbacks have been widened to increase the available floor area for luggage storage. The original wooden floors have been replaced by honeycomb aluminium floors covered in rubber matting.

The J door has a receptacle for a fire extinguisher, this is normally accessible from the saloon but can be accessed from the cab when the emergency equipment pod has been opened.

The original R and S doors have been repainted and refitted, the original door lock has been retained.

The existing passenger side ventilation units, two units on each side of each seat bay, have been replaced.

Each car has been fitted with seven ceiling fans which are thermostatically controlled, the same thermostat also controls the saloon heating and is automatically switched on by the driver. Access to the ceiling fans is obtained by removing the grille/panel assembly covering each fan.

4.3 SEATS

The general aim was to create a saloon seat of high fire resistance, low weight, ease of under seat access, aesthetically pleasing, comfortable and durable. In parallel with this a series of perch seats, an instructor's seat and a new driver's seat were required. The latter to be able to fold up for cab access and provide longitudinal and vertical sliding adjustment. Whilst giving a more comfortable seat for driver's of various heights.

Saloon seats are constructed from:- A glass reinforced phenolic resin with aluminium reinforcements bonded into position. The armrest loads were partly transmitted via the aluminium parts, via the bonding to the phenolic. The key feature being the integrity of the bonding. Gas struts counter balanced the seats for under seat access.

Drivers seat constructed from:- A steel structure to withstand rough treatment and incorporating nylatron GS runners to give low friction, low maintenance & long life sliding members. Vertical adjustment is controlled by a locking type gas strut operated by a convenient lever integral with the seat frame.

Saloon, instructor's and driver's seat were subject to the LCD 3 structural load tests and passed these satisfactorily.

5.0 ELECTRONIC PASSENGER COMMUNICATION SYSTEMS

During refurbishment the original Audio Communications system used by the driver to address the passengers and the original Destination Blind and Train Number Indicator have been removed and replaced by a Passenger Information System which provides both audio and visual information to the passengers. The PIS is made up of two sub-systems, and Audio communications system and the Visual Display system.

The Audio Communications system provides the following functions:-

- Cab to Cab
- Public Address
- Station Announcements
- Passenger Talkback
- Train Radio Interface

The visual Display system provides the following functions:-
• Train Number indication at front and rear of train
• Destination indication at front and rear of train
• Visual display of passenger information in the saloons

The Audio Communications system communicates with the Visual Display system via an RS422 serial data link. This allows the passage of random message and walktest triggers from the Audio Communications system to the Visual Display system and the passage of route message triggers from the Visual Display system to the Audio Communications system.

The driver can make 'random' announcements by entering a code on the keypad of the Cab Audio Display Panel. Where applicable a visual equivalent of these announcements is displayed in the saloons. The Visual Display System the Train's route and will trigger station announcements, both audio and visual, at the appropriate point. Provision has been made in both systems for an RS485 serial data link to a future Track-Train Data Link. Although provision has been made in the refurbishment of the train for an electrical interface with the Track-Train Data Link the hardware is an LUL project and is not yet available.

The original train radio equipment is retained in the refurbished trains. The Audio Communications system interfaces with it to give the driver the facility to speak to the line controller via his handset. The Visual Display system interfaces with the train radio to provide the train number for train identification. (The original thumbwheel switches are removed during refurbishment.) In an emergency the line controller can make PA announcements to the passengers via the Audio Communications system through the train radio.

Both the Audio communications system and the Visual Display system are software driven. Software updates for the Audio Communications system are installed by fitting a new 'Flashcard' into the Cab Master Transmitter doing this in one cab will automatically update the software in all Cab Audio Transmitters and Saloon Receiving Amplifiers on the train. While fitting a flashcard on the train is possible it is a job which should only be undertaken in a clean environment. Software updates can be installed directly into the Destination Indicators via the data socket which is mounted on the inside of the rear panel. After removal of the retaining screws the panel can be hinged down to access the socket. The software can then be downloaded from a laptop computer acting as a 'fillgun'.

6.0 CAB AIR CONDITIONING & SALOON VENTILATION

This system is required to achieve a low ambience noise within the cab environment. The drivers cab air conditioning unit delivers forced, temperature controlled, air to the drivers cab, - 70% recycled, 30% fresh (approximately.)

The system comprises a refrigeration unit, fluid unit, fan coil unit, pressure gauge and temperature control pane. The refrigeration unit is self contained and controlled by circuits within the fluid unit. The fluid unit comprises a reservoir, immersion heater, pump, system control electronics and relays.

The system is charged with a mix of water and glycol which is pumped from the fluid unit to a heat exchanger in the refrigeration unit and up to the fan coil unit before being returned to the fluid unit. The fan coil unit is mounted in the cab roof void and the refrigeration and fluid modules are suspended from the vehicle underframe. The temperature control panel and system pressure gauge are mounted on the J wall. Cab temperature is monitored by a sensor mounted in the fan coil unit (located at the J wall side.) The temperature control switch is located on the cab back wall.

When the system is switched on, logic circuits within the temperature control panel monitor the temperature difference between the cab setting and cab ambient temperature. This automatically sets the system into one of four modes in order to achieve the set temperature. These modes are:

1) Heating and low speed fan.   3) Cooling and low speed fan.
2) Low speed fan only.   4) Cooling and high speed fan.

Monitoring of temperature difference is continuous and the logic stages always select a mode appropriate to the magnitude and direction of the temperature difference.

To provide adequate ventilation in the passenger saloons, seven Elta fans are fitted in the roof space along the centre line of the vehicle. Because of their positioning and the need to provide adequate head room, the design had to be compact, with air being drawn vertically upwards into the fan and discharging radially via water separators through slots in the vehicle roof. The mixed flow type fan used provides good flow conditions against the pressure losses of the vehicle, while noise levels and power consumptions are minimal. All seven fans are run together under normal circumstances but two only provide the required flow in emergency conditions when only battery power is available. To conserve power under this and general running, the motors are especially designed to operate at reduced speeds when required, this being achieved by voltage control.

7.0   FITTING OF NEW WINDSCREENS & BODYEND WINDOWS

New impact resistant glass windscreens are fitted as a nominal 'straight' replacement for existing. The glazing system is however different in that the screen is direct bonded into the carrying frame.

8.0   PASSENGER DOORS

The bodyside sliding door system consists of mechanically guided doors powered pneumatically, and controlled electro-pneumatically under normal circumstances. A limited number of doors are capable of being operated solely pneumatically for emergency access/egress purposes, whilst a further limited number of doors are capable of being opened mechanically under air-off conditions, for crew/maintenance staff access purposes.

Under normal operation the majority of doors can be closed by 'selective close' electro pneumatic door valves in order to conserve car heat & to reduce draughts during extended platform waiting times, in inclement weather, above ground.

Due to the fact that the ratio of open cylinder to close cylinder areas is not conducive to achieving the required dynamic door performances it has been necessary to employ differential choking on both open and close cylinder lines. In the close cylinder line this is achieved using an existing fitting incorporating a fixed orifice of 1 mm.

In the open cylinder line the choke is provided by a 4.1 mm diameter choke orifice at the selective close electro-pneumatic door valve exhaust port, and a 2.9 mm diameter choke orifice at the normal (single engine) electro-pneumatic door valve exhaust port.

The guidance mechanism has been replaced to improve the load bearing and reliability characteristics of the door system. The door upper tracking has been re-designed to eliminate the current two point contact geometry of the existing vee section track in favour of a single point/line contact radiused track. Due to the geometry of the door the centre of gravity is significantly outboard of the top track therefore the track has been inclined at 12° in order to align static forces through the perpendicular axis of the bearing.

The saloon door bottom guidance consists of a mushroom type roller mounted via a journal pin in a bracket. The rollers run on stainless steel counter face strips and infill plates mounted to doorleaves. The strips engage with vee blocks at the end of the door closing strokes both at single and double leaf doorways. The counterface strips incorporate a 'hook' feature for carbody roll over, strength requirements.

9.   CCTV

During the refurbishment of the Piccadilly Line fleet the facility has been added for closed circuit television monitoring of each saloon. The system consists of two monochrome video cameras, and a video cassette recorder for each saloon being monitored. The system is in operation whenever a cab is opened up. The tapes run continuously, automatically rewinding when the end of the tape is reached. The system is powered via the loadshed relay, so that the CCTV is loadshed when traction power is lost. The cameras are fitted above the R and S doors (J door on DM and UDM cars,) looking down the length of the saloon so that between them they cover the entire saloon area.

10.   UNDERFRAME ARC BARRIERS

Underframe arc barriers are part of a package of modifications introduced to provide protection from unfused power arcing. In summary the total system comprises and protects as follows:-

Arc Barriers

a)   Withstand full arc until snubbing device comes into effect.

b)   To prevent penetration of arc into body structure.

Snubber
a) Arc extinguisher.
b) While are barrier protects vehicle, snubber prevents progression of arc.

Clearance Barrier
- Prevents arc ignition.

Arc Splash Barrier
- Prevents incandescent erosion.

Fire Barrier
- Prevents heat damage.

The arc barrier materials are generally based around Miconite EM72 both in hand lay and dough moulding compound, depending on the physical configuration of the installation.

11.0 DETRAINMENT

A detrainment device is provided at the cab positions to detrain passengers to the track in case of emergency, this replaces the existing wooden ladder. It is simple in operation and originally designed to be deployed by an un-assisted passenger. During value engineering this was changed to being operated by the driver. The angle and depth of the detrainment stairs corresponds to domestic stairs with a handrail on each side.

The whole device is constructed of Stainless Steel and the stair section is housed in a flat box which replaces the conventional sill plate and is partially recessed into the cab floor, the front, open, edge (from which the stair assembly deploys) being in the same place as the front of the conventional sill plate.

The driving end door may be used for crew access to the cab as at present, the retracted handrails being used in the same way as the existing commode handles to aid the process.

To deploy the device, the driver exits, via the 'M' door and releases two 'Gedore key' latches on either side of the 'M' door. On release of the second latch, a spring mechanism ejects the device approximately 75 mm, allowing the driver to hold the stair assembly and pull outwards. The stair assembly will then slide from within the cab floor cavity outwards, until fully extended, at which point the treads drop vertically forming a step arrangement with handrails from the cab to track level. Passengers are then able to detrain via this staircase to the track

A high level detrainment light on the cab front and two low level detrainment lights on the headstock, operated by a sealed switch in the cab, illuminate both the device itself and the area in front of the device to aid in the safe movement of passengers.

12.0 STATIC CONVERTER

The static converter is used to supply power for saloon ventilation and air conditioning for operators cabs. Static converters are used to augment the existing machine alternator set due to the following reasons:

- High reliability in excess of 20,000 hours. • Less maintenance cost.

The trailer car of each 3 car unit is fitted with a static converter. The static converter utilises its 630 Volt supply from the driving car or a UDM. The 630 Volts DC track supply is collected by the positive and negative shoe gear and fed through to the input terminals of the static converter.

The output terminals and the converter supply the cab air conditioning and saloon ventilation loads. The output is a nominal 240 Volts RMS at 50 Hz power.

LUL have imposed strict safety targets to ensure the electromagnetic compatibility (EMC) of the converter with its environment. EMC study played a major part of the design stage as the static converter has to meet the stringent signalling requirement. There should be no wrong side failure within 100 years period. The EMC design has been made difficult having the converter switching at high frequency in excess 10kHz and the notorious 33.3 Hz track circuit. The EMC design has to ensure the line current harmonic does not exceed current limit of the low frequency track circuits under all operating and failure conditions. Such harmonics could, if they occur at certain frequencies and of sufficient magnitude, interfere with the signalling system with potentially disastrous consequences. The static converter has to ensure that no excessive level of magnetic fields emitting from the converter can adversely affect the safety of the signalling system and passengers with pace makers. The EMC sensitive components have to go through full FMECA analysis to ensure that the static converter would meet the MTWSBF 10^9 hours in the calculation. Full component type tests are carried out to make sure that these components meet the EMC design Criteria.

This will be the first static converter to run over the 33.3 Hz signalling track circuit in London Underground History, this is an achievement from the EMC design point of view.

Authors:
M. D. Thomas, Bombardier Prorail Ltd.
D. Morphew, London Underground Ltd.

C511/20/089/96

Combating graffiti on rail vehicles

D B GRANDY BSc, MSc
ABB Daimler-Benz Transportation (Rolling Stock) Limited, Derby, UK

ABB Daimler Benz Transportation (Rolling Stock) Ltd. (Adtranz) has carried out an extensive laboratory study of the anti-graffiti performance of a considerable number of removal chemicals and barrier materials on a range of substrates. The results of this work, together with experience gained in service, form the basis of this paper.

The problems posed by graffiti on the exterior and interior surfaces of railway rolling stock are reviewed. The various types of graffiti and the range of materials affected and finishes subject to them are considered. An assessment is made of the pernicious effects of graffiti and those combinations of graffiti agent and surface which are most problematic.

Methods of combating graffiti are covered for both the vehicle designer and the train operator. Design alternatives are reviewed from a number of perspectives, including materials selection, interiors detailing and psychological considerations. The options open to the operator are assessed and recommendations are made as to the best use of removal chemicals and barrier layers.

1. THE GRAFFITI PROBLEM

Graffiti, from the mindless doodle on the back of a seat, to the well planned assault on the exterior of a train by a skilled artist, cost train operators hundreds of thousands of pounds a year. The considerable increase in life cycle costs arising from the problem is a result of the following factors:

(i) The high cost of cleaning the train, both in terms of personnel and materials.

(ii) The cost of specifying and using graffiti-resistant materials on new and refurbished trains.

(iii) Loss of revenue - while a particular train is withdrawn from service.
 - vandalism is a deterrent to would-be passengers.

(iv) The cost of stripping and painting the exterior of a vehicle or replacing damaged interior trim, in extreme cases.

(v) A reduction in the life expectancy of affected parts.

(vi) The cost of extra security measures taken.

2. TYPES OF GRAFFITI

Almost any method or medium for marking a surface can be, and has been, used for applying graffiti, so it is difficult to legislate for all eventualities. However, the main types of graffiti agent are as follows:

(i)   Aerosol spray paints.

(ii)   Felt-tip marker pens.

(iii) Indelible dyes, e.g. shoe dye.

(iv) Others, including ball-point pen and waxy substances such as lipstick and crayons.

(v)   Engraving.

2.1 Aerosol spray paints

The most widely available aerosol spray paints are cellulose-based; as these form the bulk of those supplied to the D.I.Y. automotive repair market. Fortunately, they are generally the least problematic when used on most hard decorative surfaces, providing the substrate is not attacked by the carrier solvent. This is because they are 'reversible', drying solely through the evaporation of the solvent in which they will readily re-dissolve. However, aerosol paints whose drying mechanism involves an irreversible chemical reaction will not simply re-dissolve in their solvent, but usually require the use of a more powerful proprietary remover. It is usually impossible to differentiate between different types of paint once they are dry, without carrying out sophisticated tests.

2.2 Felt-tip marker pens

Solvent-based permanent marker pens are more difficult to treat than cellulose spray paints, although the surface area the resulting graffiti cover is usually considerably less. The severity of the attack for a given substrate is crucially dependent on the make and colour of the pen; black and red being particularly difficult to remove. Although most pens now use alcohols rather than more powerful glycol ethers or aromatic hydrocarbon solvents, it would seem that more subtle factors have an important role to play in determining the degree of penetration into a particular surface. These include the type and concentration of dyestuff used, the particular alcohol, other ingredients and the relative proportions of the various constituents. It is often the lower edges of a graffito that are the most difficult to remove and which most often leave an indelible ghost or shadow. This is because solvent concentrates here under the action of gravity and therefore takes longer to evaporate, giving more time for the substrate to be attacked.

2.3 Ball-point pens

The inks used in ball-point pens are similar to those used in marker pens, but have a higher resin and dyestuff content. Ball-point pen is more likely to be responsible for 'unpremeditated' graffiti encountered in relatively hidden locations, such as toilet cubicles and on seat back tables, rather than as large scale graffiti. There is often an added problem caused by their scoring action.

2.4 Wax-based substances

Wax crayons can vary considerably in their formulation and it is the lower quality ones containing higher proportions of additives and fillers, such as china clay, which cause most problems. Lipsticks are made from vegetable-based wax with added pigment and are usually relatively straightforward to remove.

2.5 Dyes and polishes

Products such as shoe dyes are specifically formulated to penetrate into materials and so utilise powerful solvent mixtures and highly concentrated dyestuffs. This means that their removal presents a serious challenge to the train operator.

It is not always possible even to identify what has been used to apply graffiti; as large refillable pens can be loaded with a 'designer cocktail' of substances. This can make removal especially difficult.

2.6 Engraving

Engraving is becoming an increasingly serious issue, particularly on windows where it can badly affect mechanical properties, notably impact resistance. It presents an entirely different type of problem; one which is very difficult to combat, both in terms of prevention or repair. The UV-curable acrylic resins used to repair minor cracks in car windscreens provide a possible method of treatment, but more work is required in this area.

3. MANAGING THE GRAFFITI PROBLEM

There are a number of ways of managing the graffiti problem and the operator is well advised to use all of them. These may be categorised as either methods of preventing or reducing the incidence of graffiti, or as methods of treating or reducing the seriousness of attacks when they occur. As with most problems, prevention is usually better than treatment of the consequences, although it would be prohibitively expensive to eradicate graffiti altogether. This means that the most cost effective balance, encompassing both prevention and treatment, must be devised.

Figure 1. Combating Graffiti

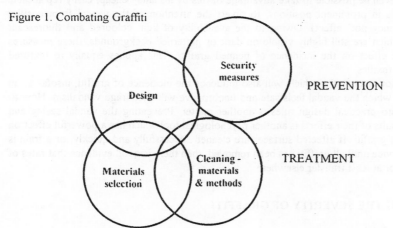

This illustrates the interdependence of the measures that can be taken to prevent or treat graffiti - vehicle design and security precautions affect the incidence of vandalism, whereas the selection of 'graffiti resistant' materials and the use of cleaning procedures can both be classified as ways of treating the results. Design and materials selection are, of course, intimately related, whilst the materials used will have a crucial bearing on the efficacy of any graffiti-removal process. Less obvious are the effects that design (for example, interior layout) will have on the ease of cleaning, the success of security measures taken (is it possible to hide from the security cameras?) and on the frequency of attack.

4. REDUCING THE OCCURRENCE OF GRAFFITI

4.1 Security measures and housekeeping

Increased security measures at a stabling yard (high fences, razor wire, floodlighting, patrols, video-surveillance) will reduce the incidence of severe attack occurring on the exterior of train. Installation of closed circuit television, increased staffing levels and successful pursuit and prosecution of offenders will all help reduce the levels of graffiti attack on vehicles in passenger service. Most of these methods are expensive and an analysis must be carried out to determine the cost effectiveness of a particular measure.

Environments which show obvious signs of neglect tend to attract vandalism, which in turn attracts further vandalism, whereas an environment that is kept clean and tidy and displays obvious signs of 'ownership' is less likely to suffer. Anti-graffiti measures should therefore form part of an efficient routine cleaning operation.

4.2 Design

The use of elements of interior design to reduce the incidence of graffiti attack is a relatively recent, and as yet little-studied, concept. It does overlap with the area of materials selection, in that different finishes, colours, patterns and textures do attract varying amounts of graffiti. Large areas of light colour seem especially irresistible to the vandal (the 'white-board' effect), whereas bolder colours and patterns appeal considerably less, because any graffiti will be less visible. It may even be possible to take advantage of this by installing cheap, easily replaceable panels or decals in prominent positions, to divert the attention of the vandal from costlier targets (the 'honey pot' effect). Owing to the availability of light coloured and fluorescent spray paints, which are still highly visible on dark or patterned backgrounds, these measures will have more effect on the incidence of penned graffiti. This applies equally to textured surfaces and to textiles.

The interior layout of a vehicle will also influence the incidence of graffiti, insofar as an environment in which the vandal feels safe and unobtrusive will encourage vandalism. How to translate this into practical design rules is another matter. Preventing the vandal seeing and enjoying the results of their efforts is another psychological tool which has a powerful effect on the incidence of graffiti. If affected surfaces are cleaned successfully and quickly, or a train is taken out of service until all graffiti has been removed, then there is good evidence that rates of vandalism fall, or at least transfer elsewhere.

5. REDUCING THE SEVERITY OF GRAFFITI

5.1 Materials selection

A measure of the 'graffiti-resistance' of a material is its resistance to solvents, not just those used in the graffiti medium itself, but those employed in any cleaning chemicals which may be used subsequently. When a surface is sprayed with paint or written on with a felt marker pen, it is the interaction of the carrier solvent with the substrate that determines the seriousness of the attack. If the carrier solvent also acts as a solvent for the substrate material, then local swelling or dissolution will occur and the solvent will penetrate into the material, carrying with it the other ingredients of the paint or ink. Once the solvent has evaporated, the paint or ink is left within the surface of the substrate making it very difficult to remove, without further (permanent) damage to the component. Even with the best cleaning chemicals, 'ghosting' or 'shadowing' usually results.

Organic solvents may be categorised as polar or non-polar. Polar molecules contain atoms or functional groups which either attract or repel electrons in a chemical bond. Polar bonds include C-X (where X is a halogen), C-CN, C=N, C=O, C-OH and examples of polar solvents are:

$$
\underset{\text{1,1,1 Trichloroethane}}{Cl-\underset{\underset{Cl}{|}}{\overset{\overset{Cl}{|}}{C}}CH_3}
\qquad
\underset{\text{Acetone}}{\overset{CH_3}{\underset{CH_3}{\diagdown}}C=O}
\qquad
\underset{\text{2 Methoxypropanol}}{CH_3\,\underset{\underset{OCH_3}{|}}{CH}\,CH_2OH}
$$

Non-polar organic solvents are usually low molecular weight aliphatic or aromatic hydrocarbons e.g. hexane and xylene. It is generally the case that materials made up of polar molecules will be resistant to attack from non-polar solvents and *vice versa*.

Most finish materials used on a train are polymeric: thermosetting resins are used in coatings, laminates and mouldings; thermoplastics are used in sheet form and as mouldings; elastomers can appear in the form of trim extrusions, floor coverings and gangway diaphragms. The solvent resistance of thermosetting or cross-linked polymers, including elastomers, is determined partly by the chemical structure of the molecule, and partly by the degree of cross-linking. Those containing many polar functional groups have good resistance to hydrocarbon solvents and, in general, the higher the degree of cross-linking, the better the solvent resistance. However, it is not only the basic structure of the polymer that affects graffiti resistance, but how well any additives, such as mineral flame retardants, are incorporated. If there are uncoated particles at the surface of, say, a polyester gel coat, then these can effectively act as pores, allowing solvent penetration to occur. The solvent resistance of a thermoplastic is determined to a large extent by its ability to crystallise; highly crystalline polymers such as polyethylene, polyamides and polyacetal have good chemical resistance, whereas amorphous plastics such as polystyrene, poly(methyl methacrylate) and, to a certain extent polycarbonate, have relatively poor chemical resistance.

Graffiti resistance is of course only one of the many criteria that must be satisfied when selecting materials for use on rail vehicles. In many applications it may not always be possible to use a material with the best possible anti-graffiti performance, without compromising other properties to an unacceptable degree. Examples of this are: very highly cross-linked thermosetting resins, including those used in coatings, which are often too brittle to be practicable; halogen-containing materials, such as poly(vinyl chloride) and poly(tetrafluoroethylene), which produce highly toxic gases on combustion; high pressure melamine laminates, which cannot be formed into curved panels; and highly glossy coatings, which may not meet visual requirements.

Vitreous enamel and some ceramic coatings have excellent chemical resistance, but their use is somewhat limited by their cost, weight and mechanical properties, particularly their vulnerability to impact damage.

Bare metallic finishes are undamaged by solvents and are therefore usually easy to clean, although aluminium alloy surfaces will gradually succumb to ghosting after repeated cleaning with, for example, phosphoric acid based solutions. This is because the oxide layer becomes increasingly porous with time. This problem can be alleviated by using anodized and sealed material. The use, however, of substantial areas of bare metal is usually prevented by visual considerations, weight, processing difficulties (e.g. producing complicated shapes) and high thermal conductivity. Where bare metal is acceptable, trim panels made from brushed stainless steel, for example, offer excellent anti-graffiti performance.

Table 1 The graffiti resistance of common finishing materials

Material	Solvent resistance	Comments
Thermosetting Resins		
Unsaturated Polyester	G	Usually gel coat
Phenol Formaldehyde	-	Requires coating
Melamine Formaldehyde	VG	Decorative laminates
Epoxide	G	Structural composites
Vinyl ester	VG	"
Acrylic	G	e.g. seat shells
Thermoplastics		
amorphous	Generally M - P	
semi-crystalline	Generally VG - E	
Organic coatings		**High gloss usually best**
Alkyds	M	Used on older stock
2k Polyurethane	VG - G	Clear lacquer usually best
2k Epoxy	G	Prone to yellowing in UV
Epoxy powder	G	"
Polyester powder	G - VG	
Polyurethane powder	VG	
Thermoplastic		e.g. nylon dip
Vitreous enamel	E	Brittle, heavy, costly
Metals		
Aluminium alloys	G	Ghosting can occur
Stainless steels	E	

E = excellent, VG = very good, G = good, M = moderate, P = poor

5.2 The use of barrier materials

There are two types of barrier material:
(i) Self-adhesive polymer films.
(ii) Spray or brush-on sacrificial coatings

Films can be clear or self-coloured and range in thickness from $20\mu m$ to $80\mu m$, the most common materials being PVC and polyester. Acrylic adhesives are often used, making them very difficult to peel off (they can usually be removed only with the help of a heat gun). They can be applied either as part of the manufacturing operation or retrofitted to vehicles in service. Most are suitable only for flat or single-curved components and are best applied floated on a detergent solution to eliminate bubbles and creases. Care must be taken to avoid exposed edges where possible, as these pick up dirt and attract the attention of the vandal, so, where possible, edges should be concealed under adjacent items of trim. This is done more easily as part of the production process, but it is important to ensure that the film can be replaced in service with the minimum of difficulty, should it become damaged.

Sacrificial coatings contain low surface energy components such as PTFE or silicones which are left as a waxy deposit on the treated substrate. They prevent the graffiti agent penetrating to the substrate and are subsequently removed, along with the graffiti, by a suitable cleaning agent. Another type of coating utilises drying inhibitors, so any paint in contact with it remains tacky for a considerable time. In this state it can, in theory, be removed more easily. Surfaces must be re-treated at regular intervals to maintain the effectiveness of these coatings.

Self-adhesive films are expensive and their installation is a relatively skilled operation. When applied to coated surfaces, they increase substantially the amount of organic material present and therefore must be subjected to all the applicable fire and smoke tests for each substrate treated. Although fire performance is not usually a problem for most films, smoke emission limits for underground vehicles are more difficult to achieve. The use of PVC films is usually prohibited, on UK vehicles at least. For these reasons it may well be prudent to use polyester films strategically in especially vulnerable areas, to minimise the extra cost and potential smoke load. The performance of some of these films is excellent and their use should be considered seriously.

Liquid sacrificial coatings cost less but are much less effective than self-adhesive films. Their use is less dependent on skilled personnel, but care must be taken to avoid contaminating any textile surfaces adjacent to treated areas. The use of formulations containing even small amounts of PTFE is usually prohibited on UK vehicles, because of the threat of hydrogen fluoride evolution during combustion.

It is possible that the use of self-healing coatings, developed to minimise spall from windows during an impact, will make glass less vulnerable to engraving, although such coatings will not afford protection against a determined attack. They have as yet built-up little in the way of service record and will inevitably have a detrimental effect on fire performance. The cost of such coatings is considerable.

6. GRAFFITI REMOVAL

There are many proprietary graffiti removal chemicals on the market, ranging from relatively mild alcohol-based formulations supplied by pen manufacturers, to extremely aggressive products containing dichloromethane intended for use on masonry and other non-organic surfaces. It is vital, therefore, that before a cleaning chemical is used on a vehicle, it is tested thoroughly on all applicable finish materials, a detailed procedure formulated and all relevant personnel are trained in its use. Failure to do so could lead to the indiscriminate use of aggressive chemicals, causing even costlier damage to a vehicle.

Common active ingredients in removers used on rail vehicles include N-methyl-2-pyrrolidone (NMP), which is a somewhat less aggressive paint stripper than dichloromethane, methoxypropanol (an alcohol often used in inks and dyes) and propylene glycol methyl ether. Products are often available in both liquid (spray or wipe-on) and gel forms. Gels have traditionally been formulated for use on porous materials, because more mobile solvents can often soak into the substrate, carrying the dye or pigment with them. They can also be useful on smooth decorative finishes, because they are less messy than liquids, but it is often tempting to prolong exposure times on particularly stubborn graffiti, leading to damage of the substrate. For this reason, it is important that in any cleaning procedure, careful attention is paid to maximum safe exposure times. It is more difficult to overexpose surfaces when using liquid cleaners, but care must be taken to avoid the contamination of adjacent areas with runs or splashes of dye-loaded remover.

If ghosting remains after the cleaning operation, it is usually futile and indeed potentially damaging, to continue treatment with the same remover. In these instances, a bleach solution is often effective.

Waxy agents, such as lipstick and crayon, can often be removed using vegetable oil or white spirit, which are less damaging to organic substrates than powerful graffiti removers. Detergent solutions can often selectively dissolve the wax component when used on this type of graffito. This leaves behind the pigment, making treatment more difficult.

It is essential to treat any graffiti as soon as possible after the attack has occurred, in order to minimise reaction with the substrate. This makes efficient inspection, reporting and cleaning procedures vital. Adequate training of personnel, not least in the health and safety precautions necessary, is a prerequisite to any successful cleaning operation. It is usually the case for a particular vehicle, that one removal chemical will not be successful for all combinations of graffiti and substrate. Consequently, it is advisable to employ a battery of products, ranging from the relatively benign pure alcohols to the paint stripper type, so that as many permutations as possible are provided for. This has the advantage that the use of aggressive chemicals is avoided when a simple alcohol-wipe will suffice, but makes procedures more complicated and cleaning more of a skilled job. It is always good practice to treat a small, relatively unobtrusive area first to ensure that a particular remover will work without damaging the affected component. In order to minimise the damage to treated surfaces, the use of harsh abrasive pads should be avoided.

In extreme cases, where most of the exterior of a vehicle has been covered in graffiti, it may well be more cost effective to carry out a complete re-spray, after stripping the bodyshell back to bare metal. This will usually require the use of a specialist surface preparation operation, such as plastic bead blasting. Simply painting over a graffiti-contaminated surface can lead to problems with bleed-through and will affect the fire performance of the coating system by increasing the dry film build. Another option is selective removal of the contaminated top coat, using a less harsh solid carbon dioxide blast, followed by re-activation of the primer. Basecoat and clear lacquer, or the direct colour finishing coats can then be applied. However, this requires the use of specialised carbon dioxide stripping equipment.

7. GRAFFITI TESTING

Testing the graffiti resistance of a material or the efficacy of a cleaning product can be an involved process. Firstly, the goal of the test programme must be well defined: is the aim to identify a better cleaning procedure for existing materials or is it to identify new materials having improved graffiti resistance? Secondly, a performance criterion must be defined which removes the observer's subjectivity as much as possible. Thirdly, the number of variables must be kept to a minimum, in order to restrict the amount of testing to within reasonable limits. As the possible variables include the substrate material, its colour and gloss level, the graffiti agent, its colour and exposure time, the type of remover, its exposure time and number of application and removal cycles, careful experimental design is essential.

The test programme conducted at Adtranz was designed to achieve the following goals:

(i) To identify both improved cleaning products and procedures together with possible barrier materials for existing vehicles.

(ii) To evaluate alternative coatings for use on new or refurbished stock.

For this reason, the programme could be divided conveniently into two parts; the first being the appraisal of a considerable number of proprietary and generic removal chemicals on on coatings currently in use on Adtranz vehicles (polyester powders and two pack polyurethane paints). Barrier films and coatings were also tested at this stage. The second part involved the evaluation of a number of alternative coatings.

Black, red and blue spray paints and permanent marker pens and black shoe dye were used as the graffiti agents. Each graffito was left to condition for twenty four hours before the remover was applied, left for a fixed time and cleaned-off in accordance with the manufacturer's instructions. Performance was then rated on a scale of 1-5; 5 corresponding to complete removal, 4 to slight ghosting and so on. The procedure was then repeated on the same area until a substantial deterioration in effectiveness occurred. The number of cycles to reach this stage was recorded.

The principal findings of the Adtranz test programme are summarised below:

- The worst graffiti agent was black shoe dye, followed by black, red and then blue marker pen.

- Cellulose-based spray paints could usually be removed quite successfully from most substrates for a number of cycles, using a wide range of proprietary removers.

- No single remover was identified which performed best in all situations. It was therefore concluded that it would be prudent to approve the use of an NMP-based product and one other proprietary mixture, in both spray and gel form, in order to best treat most combinations of material and graffiti.

- Pure alcohol solvents were found to remove some marker pen graffiti more successfully than the proprietary removal chemicals, with much less risk of damaging the coating.

- The coatings which performed best were polyurethane powders and two-pack lacquers. Gloss level was found to have a marginal effect. Clear-over-base systems were shown to be slightly better than direct gloss. The only coating from which shoe dye could be removed with any degree of success was a proprietary gloss powder.

- Sacrificial barrier coatings were reasonably effective against spray paints, but were found to afford little extra protection against pens and shoe dye.

- Self-adhesive polyester films performed very well, with one in particular showing excellent resistance to leather dye, with no discernible deterioration even after numerous cycles. Polyester powder coated aluminium alloy panels covered in this film were subjected to fire and smoke emission tests with excellent results.

- The long-term performance of most coatings was found to be critically dependent on exposure time with all the proprietary removers. This was particularly true of concentrated gels which attacked most materials very quickly, leading to softening and loss of gloss.

- Some newer epoxy, polyester and epoxy-polyester powder coatings with promising properties were identified.

8. SUMMARY

The train specifier, designer, manufacturer, operator and maintainer all have a crucial role to play in combating graffiti. There is no panacea; a range of measures must be adopted.

When defining the level of 'graffiti resistance' a vehicle is required to meet, the different types of graffiti, removal chemicals, test methods and measures of performance must all be taken into account. It is insufficient to state simply that surfaces shall be 'graffiti-resistant'.

The interior design and layout, types of finish and even colours and textures, can all be used to create an environment which discourages vandalism. The strategic use of barrier film on vulnerable areas should be considered.

The maintainer of the train must ensure that inspection and cleaning procedures are such that graffiti are reported and treated quickly and effectively and that staff are well trained in their implementation. All practical and affordable security measures should be adopted, in order to reduce the incidence of graffiti and detect the perpetrator.

REFERENCES

1. BILLMEYER, F. W., Textbook of Polymer Science (2nd ed.), 1971, (Wiley), New York.
2. WALLACE, J., WHITEHEAD, C., Graffiti Removal and Control , 1989, (CIRIA Special Publication 71), London.

C511/20/127/96

The Value Management culture – ensuring that re-engineered products give maximum value for money

M B JEFFERYES BSc, AMIMechE, MIVM
Value Management Limited, London, UK

<u>Synopsis</u>   The planning, design and implementation of a project require effective communication and co-operation between a great many disciplines. The way is strewn with opportunities for misunderstanding, inefficiency, delay, compromise and unnecessary cost.

The Value Management (VM) process intervenes at key stages of a project to bring the responsible parties together, building them into a team, while giving them the methodology to develop their project in ways which overcome such inefficiencies. VM, best applied in intense workshop format, enables the team to maximise the Value of the project by optimising the balance between Performance, Cost and Timing.

1. INTRODUCTION

The **Value Engineering (VE)** process was developed in wartime USA and has been used extensively in the US and Japan since the mid 1950's. It has more recently been adopted in many other parts of the world, but its history in Britain has been mixed.

The present resurgence of VE in the commercial field has been stimulated by intensifying competitive pressure world-wide. Coupled with this, customers are growing more and more discerning and, knowing a good deal when they see it, are exercising their freedom of choice. To survive in this market environment, business efficiency and demonstrable value are vital.

To the Japanese, Value Engineering is a wide-spread, calculated step in their drive for world-class quality and excellence. To others, VE has been the means of survival and the key to success when faced with such competitive threat.

The public sector is also under pressure to deliver value for tax payers' money. This is captured eloquently by London Underground's goal of "Doing More for Less". The HM Treasury Guidance recommending use of Value Management was issued in January 1996.

Value Management (VM) has evolved to include a suite of studies applicable throughout the life-cycle of a project. The VM process changes its nature, adapting to the circumstances faced at different stages of a project's life.

Fundamental to all successful VM work is the Team. In theory the VM tools can be used by an individual. However, their strength lies in the way in which they harness the power of the team, while simultaneously boosting team effectiveness and cohesion.

2. WHAT IS VALUE MANAGEMENT ?

Definition:- Value Management is a structured and disciplined, team-centred problem-solving technique to ensure the optimum balance between performance, cost and time. VM is an extremely effective method for identifying and eliminating unnecessary cost.

What singles out VM from indiscriminate cost reduction is its focus not on cost reduction but value optimisation.

Definition:- Value is the reliable performance of what is required, at lowest possible cost. In other words, Value is....."What you get (of what you want) for what you pay".

In simple terms, this is defined by the equation:-

$$Value = \frac{Function}{Cost}$$

Definition:- Function is everything required by any of the customers. For example, the end user may require Performance, Appearance, Ease of Use; the Manufacturer may require ease and speed of construction; the Operator may require low energy consumption and ease of maintenance. Another customer is the Environment which has its own requirements during the construction, during the use and the disposal of the product.

Definition:- Cost is best represented by the Total Cost or the Life-Cycle Cost. VE studies almost always result in significant **cost** reduction, but not at the expense of **function** which is frequently enhanced. Cost is occasionally allowed to increase in return for a proportionately greater increase in function. Happily the majority of studies result in the win / win of increased function alongside reduced cost.

3. WHEN SHOULD VALUE MANAGEMENT BE APPLIED TO A PROJECT ?

To maximise project value, VM should be used regularly throughout the project's life, firstly to set targets, then to help achieve them. VM helps the team to prepare for, and to make, the major project decisions.

Each intervention (listed in table 1) is a brief, fast-track challenge to the project, centred on the responsible team who dissect then reassemble the project in a very structured way.

We defined Value as *"What you get (of what you want) for what you pay."*

For any project, value cannot be determined until you can answer the *" what you want"* question. This is the purpose of the first Value Planning intervention - VP1.

VP1 occurs at the start of the project, facilitating the definition of objectives and concept. It assembles the key project stakeholders and develops the detailed project objectives, through consensus. Equally important is achievement of a <u>shared understanding</u> of those objectives so that future efforts of different branches of the team are directed towards the same goals.

The output is often a number of alternative schemes, all of which are then explored through feasibility studies.

VP2 occurs after the feasibility studies, giving the team a mechanism for selecting the option which delivers highest value in relation to the established project objectives. It builds the decision based on a series of judgements, many of which are subjective and often emotional, leading to team consensus in a structured and accountable method. Once a preferred option is selected, the team then seek further value enhancements. The output is one agreed scheme which is subsequently taken through outline design.

The Value Planning stages have addressed **"What"** is required of the project. Value Engineering now determines **"How"** the project will be delivered to satisfy the established objectives.

VE1 then VE2 come during the Outline Design and then the Detail Design periods. This is traditional Value Engineering, maximising value by optimising the relationship between high Function and low Cost - ie. *"What you get for what you pay"*

VEn studies are called spontaneously as required to deliver fast, agreed solutions to design or construction problems which arise during the project implementation phase.

VP3 is sometimes run as a project post-mortem, what was good or bad, to learn lessons for the next project.

VA (Value Analysis) studies may be necessary after project implementation if concerns are identified or if significant ongoing costs indicate opportunities for value improvement.

These different interventions in the project life are shown in table 1 - and in the graph of VM Timing versus Savings Potential over the Lifetime of the Project.

Table 1. Different VM study interventions at different project stages

Project Stage	Value Objective	V.M. Service
Project Definition & Feasibility	Consensus on highly defined Project Strategy	Value Planning *VP1*
Conceptual Design	Guarantee of Best Option	Value Planning *VP2*
Preliminary Design & Engineering	Maximum Value in the Overall System Design	Value Engineering *VE1*
Detailed Design & Engineering	Maximum Value in the Detail Specifications	Value Engineering *VE2*
Implementation	Rapid resolution of problems as required	Value Engineering *VEn*
Commissioning & Handover	Objectives Met, Efficiency, No Surprises	Results of Team Building+Partnering
Post-Project Evaluation	Lessons learned for next time	Value Management *VP3*
In operation	Improved Value in operation Resolution of concerns in use	Value Analysis *VA*

4. SAVING POTENTIAL THROUGH THE LIFE OF A PROJECT.

The graph below shows that savings are at their greatest in the early stages. Later on, certain decisions become locked in, limiting the scope for change. At the same time, costs of change progressively increase as some work must be repeated and as special actions are required to implement changes within the reducing time available. Nevertheless, VE studies during the implementation phase can still produce significant benefits, both in savings achieved and in avoidance of cost and time overruns.

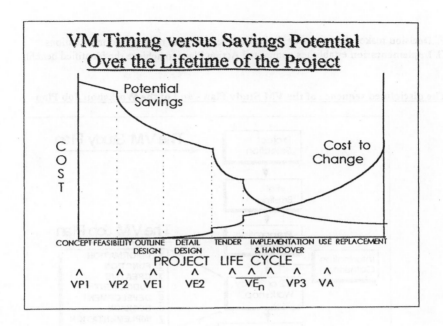

VM Timing versus Savings Potential Over the Lifetime of the Project

Potential Savings

COST

Cost to Change

CONCEPT FEASIBILITY OUTLINE   DETAIL   TENDER   IMPLEMENTATION   USE   REPLACEMENT
                      DESIGN   DESIGN   & HANDOVER

PROJECT LIFE CYCLE

∧   ∧   ∧   ∧   ∧ ∧ ∧   ∧
VP1   VP2   VE1   VE2 $\overline{VE_n}$ VP3   VA

5. HOW DOES VALUE MANAGEMENT WORK ?

A Value Management study operates two systematic plans simultaneously:-

The Study Plan sets up the VM study, establishing its scope, its objectives, its groundrules, its information base, the appropriate team membership, timing and logistics. An initial briefing with 2 or 3 key project stakeholders is followed by a preparation meeting with perhaps 6 or 7, then a period of careful data collection. The VM Workshop follows, attended by the full team. After the Workshop, a draft report is presented for the team's comment so that a final and fully agreed report can be published. The report records the Workshop proceedings, conclusions and the team's committed action plan to achieve the identified benefits.

The Workshop itself follows a tried and tested 7-step Workshop Job Plan.

A. **Information** is shared in a series of brief presentations
B. **Function** examines what the subject <u>DOES</u> and <u>MUST DO</u>, <u>NOT what it IS</u>. It ignores the current design assumptions, looking behind them at the customer needs and wants. This interrupts the rush to find the answer, carefully examining the information and understanding the problem before trying to solve it.
C. **Creativity** then seeks many, many alternative ways (both wild and conventional) to perform these functions.
D. **Judgement** applies filters to select quickly a short list of the most promising ideas for more detailed examination.
E. **Development** determines the advantages and disadvantages of these winning ideas.

F. **Decision** makes the necessary choices and registers the team's recommendations.

G. **Implementation** establishes committed Action Plans to achieve the identified benefits.

The disciplined sequence of the VM Study Plan – and of the Workshop Job Plan

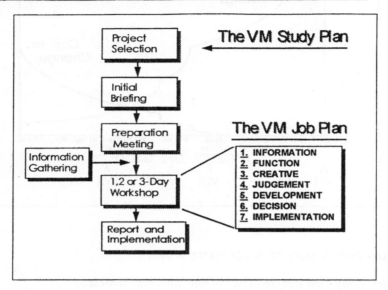

6. WHO SHOULD BE INVOLVED ?

Team membership must include all decision makers key to project delivery, all who have the greatest stakes (either for or against) in the subject matter of the study. Their participation will ensure ownership of robust decisions and commitment to implementation. Team membership will vary through the project's life and may include the following:-

- Key stakeholders - including the Project Sponsor and User
- Project Team - including Designers, Cost Control, Manufacturing & Construction
- Safety and Legislation Specialists
- Future Operators and Maintenance
- Outside Specialist knowledge - Suppliers / Contractors
- A VM Facilitator, independent of the project team.

    External expertise is of great benefit to the process - not only in the additional knowledge and experience brought to bear, but through the stimulus and challenge which 'outsiders' can bring to the resident project team.

There are a number of ways in which the outside expertise can be introduced, for example:-

a) At inception, potential contractors can be invited to participate, with nothing to gain but goodwill. Several contractors may attend on this basis, all made aware that their competitors will be present and all understanding that participation gives them no advantage in any subsequent tendering. Some organisations pay potential contractors as consultants for this involvement.

b) In design, a 2-stage tender process can incorporate a VE study. The favoured contractor from the first stage joins the team for the VE study to help refine the project design. The second stage tender then follows the VE study.

c) Post-Contract, further VE studies can produce significant benefits. In order to motivate the appointed contractor, faced with possible reductions in contract worth, incentive schemes are developed to share the resulting savings.

7. VM IN THE CONTRACTING / SUPPLY INDUSTRY

Application of VM throughout the supply chain will help contractors and suppliers achieve essential efficiencies in both design and the manufacturing / construction processes. The results lead to increased profitability in established business and, when applied during tender preparation, the opportunity for greater competitiveness.

8. WHAT DOES VALUE MANAGEMENT ACHIEVE ?

- Highly structured decision making.
- Projects correctly focused on the customer needs.
- Functional enhancements alongside cost reductions.
- Widest range of solutions considered
- Savings often 10 to 30% of project value.
- Projects properly planned and costed.
- Value optimised
- The Time to do it properly and to "Get it Right First Time"
- Team building, team understanding, consensus and co-operation.
- Committed action plans
- Smooth project implementation
- A Culture Change, with all team members sensitive to cost, value, time, the customer, each other and the business as a whole.
- Job satisfaction and pride in the project and in the business.

Examples:

» VP1 study determined objectives for the future of one Underground Line
» VP1b study for next generation of Underground Trains.
   Team of 18 generated 592 ideas, selecting those for priority attention.
» VP2 study examined 4 alternative schemes for Station Refurbishment.
   Team of 23 reached consensus that highest cost gave best value.
» VP2 study examined 4 alternative schemes for Inter-Car Barriers.
   Consensus reached, then improvements made to chosen scheme.
» VM/VE study into Rail Line routing saved 21% of project cost.
» VE1 study into Car Radio Mountings saved 44% cost and 35% weight.
» VE1 study into proposed new Trainwash saved 18% of project cost.
» VE1 study into Station Drainage improved drainage <u>and</u> saved 14% cost.
» VE2 study into a Motorway Interchange saved 19% of project cost.
» VE2 study into Train Fleet Refurbishment saved around 20% of scope studied.

DETAILED CASE STUDIES TO BE DESCRIBED:

1. VE on '73 Tube Stock (ref. presentations by Mr.David Morphew & Mr.Mel Thomas)
2. VP2 Decision on Inter-Car Barriers.

C511/21/118/96

Technical characteristics of the European RoadRailer system

P ZUPAN MSc, PhD, MSciSocMechE(Hungary)
RoadRailer (Bimodal) Limited, Bedford, UK

RoadRailer is a bimodal freight transportation system developed originally in the USA and adapted in the last five years to the European road and rail regulations and specifications, keeping the proprietary coupling design virtually unchanged.

RoadRailer's prior experience is based on 4000 trailers in service since more than 10 years in the USA and there are RoadRailers in service now in Europe, Australia, New Zealand, Thailand and India as well.

The technique of the RoadRailer system is characterised by simple terminal operation, solid durable, simple design, high level of operational safety, widespread use of standard railway components, excellent aerodynamics characteristics and good riding stability.

The first commercial service in Europe in operation since June 1995 with good reliability results.

1. INTRODUCTION

RoadRailer is a bimodal freight transportation system developed originally in the USA and adapted during the past 5 years to the European road and rail regulations.

RoadRailer's prior experience is based on 4000 semi-trailers in service in the USA during the past 8 years. Since 1990 fifteen RoadRailer protoype units have been delivered to 5 different European railways. Several units have been registered in the UK and are in process of being registered in France. The first commercial European RoadRailer service began in June 1995 between Munich and Verona with 60 trailers and 78 bogies, and another 100 trailers and 80 bogies are under delivery for the extension of the service. There are also RoadRailers in service in Australia and New Zealand and further units are under manufacture in India and Thailand.

The RoadRailer system is an ideal way for a door to door transportation logistic chain. The trailers are collected and distributed in a limited radius around the rail terminals on the road by means of normal road tractors. On the rail terminals the trailers are coupled with the railway bogies and with each other preferably to a unit train, which can have maximum 50 units in Europe (due to the 700m maximum train length). On the rail terminal the trailer is easily connected to a bogie by a road tractor then to the next trailer and coupled (Fig. 1). The road wheels are raised by air pressure and locked in place during the rail travel. The safety systems ensure that each trailer is correctly locked on the bogie. Each trailer is linked by an articulated coupling which provides a slack free connection and prevents torsional movements being transferred between trailers (Fig. 2).

All the locks can be easily checked by 'user friendly' safety systems running along the side of the train. If the trailers are not correctly secured the train emergency brake will be automatically applied preventing movement in rail mode (Fig. 3).

The fifteen various European prototypes have been tested thoroughly according to the specification of the UIC 597 Leaflet and to the specification of BR. Based on the test results, the RoadRailer system has been approved by the UIC as of 1st of July 1994.

2. THE EUROPEAN ROADRAILER SYSTEM

The European RoadRailer system is composed from:
- 13.6m long standard size trailers of curtainsider, box van, flatbed / container chassis or in the future any other usual trailer design (refrigerated box, tank). (Fig.4)

- Standard UIC Y-25 type bogies adapted to RoadRailer application. (Fig.5)

- Universal front / rear adaptors to connect the RoadRailer train to the traditional railway rolling stock (locomotives or wagons). (Fig.6)
- Middle adaptors to connect the trailers with the bogies.

3. THE CHARACTERISTICS OF THE SYSTEM

3.1. General characteristics
The RoadRailer system is characterised by a direct link without any play between two semi-trailers, from which there is a linear transmission of the traction force.

This link, a ball and socket joint type, allows any semi-trailer related movement in relation to the semi-trailers to which it is coupled.

Every semi-trailer rests at the rear on two points on the bogie adaptor unit and at the front on one point. This configuration is close to the road configuration (resting on the wheel set at the rear and on the fifth wheel of the road tractor at the front).

Linked one to another, the semi-trailers form an entire RoadRailer train.

The end adaptors are identical so that they can receive the front or the rear of a semi-trailer.

The middle adaptors are symmetrical. Their orientation is independent of the orientation of the semi-trailers. The transfer from road mode to rail mode requires coupling procedures: coupling with a road tractor, coupling of the semi-trailer on the bogie adaptor unit and finally coupling of two semi-trailers together. This procedure is possible due to the pneumatic suspension of road axle which allows the semi-trailer to move vertically.

The RoadRailer system bogie is shown in Fig.5 and the characteristics are shown in paragraph 3.4.

3.2 Characteristics of the semi-trailer
3.2.1. Box Van
The box van is a self-contained integrated aluminium body design built up from sidewalls, roof, underframe and rear and front frame (Fig.7).

The sidewall has a top and bottom side rail from large aluminium extrusion connected by the vertical posts and the aluminium side sheets. All the components are riveted together to form a loadbearing structure.

The two sidewall connected at the bottom with the front and rear underframe and a series of aluminium crossmembers between the two extruded longitudinal siderails. At the top they are connected by a series of aluminium roofbows covered by aluminium roof sheets.

The front underframe is a fabricated steel structure, containing the front coupler casting and the bolt in kingpin of the trailer.

Rear underframe high strength fabricated steel structure includes bogie lock pins, guide rails, coupler casting and linkage for coupler pin release mechanism. A second bolt-in king pin at the rear of the trailer has been positioned to mate with the rear bogie adaptor.

Coupler mechanism: low profile coupler, linkage for remote operation includes solid link between safety pin and linkage rod and spring overcentre device to hold coupler pin in the raised or lowered position (Fig.3).

The interfaces between the aluminium and steel structural members are carefully insulated to avoid electrolytic corrosion.

3.2.2. Curtainsider trailer (Fig.8)
Chassis
The main beam of the chassis is a box capable of taking the vertical and longitudinal loads (allowed tractive effort of 1000 kN).

The rear of the chassis accommodates the locking mechanisms (adaptor/semi-trailer and semi-trailer/semi-trailer) as well as the safety devices to which they are linked.

3.2.3. Axle set
The axle set is formed from three axles, a single set of tyres and air suspension with:
- Pneumatic axlelift system which lifts the axle set into rail position.
- Mechanical axle locking and visual signal which shows that the axle set is locked in rail position after the automatic dump of the air contained in the liftbags.

3.2.4. Landing legs

The legs can not be folded, 2 gears, operated by hand, with mechanical locking in raised position (rail).

3.2.5. Rear underrun bar

The rear underrun bar is mechanically locked in the road or rail position.

3.2.6. Curtainside

In accordance with UIC leaflet 597, the curtain has been submitted to a resistance test (0.3 x 28 000 kg payload applied laterally load evenly distributed on the semi-trailer's 35.1 m^{2}side curtain).

This test showed no signs of damage, nor permanent deformation of the curtain.

Due to the design of the semi-trailer, the tension of the curtain is independent of the trailer loading condition.

3.2.7 Gauge

The semi-trailers are within the UIC 505-1 and UIC 503 gauges.

3.3. Description of the connections

The connection between the two semi-trailers is provided by a tongue. The tongue is welded to the front of the semi-trailer, and is locked in the rear of the preceding semi-trailer.

A similar mechanism allows the transmission of the traction force at the level of the end adaptors.

The connection between semi-trailer and middle or end adaptor is by means of two pneumatically operated horizontal pins and two vertical pins.

The semi-trailers two pneumatic operated horizontal pins and two vertical pins ensure that the connection is maintained.

The stiff connection between front of semi-trailer and front end adaptor is maintained by the king pin and the fifth wheel of the semi-trailer.

Each locking mechanism (between 2 semi-trailers and semi-trailer and adaptor) are combined with a visual indicator showing the locked or unlocked state, easily accessible all along the train, as well as a device that immobilises the train by discharging the main brake pipe.

The RoadRailer coupling system is designed to withstand a compression and tension load of 1800 kN without any residual deformation nor breakage.

3.4 Characteristics of the bogies

The type of bogies used have the following characteristics:

Designation of the model	Y 25 LRR
Wheel base	1800 mm
Weight per axle and speed	22.5 t at 100 km/h
	20.0 t at 120 km/h
(*) Height of the spherical pivot/rail	875 mm
(*) Height of the flexible sidebearers/rail	905 mm

(*) These two measurements are given under a load of 10 t on rail.

Double block clasp brake, integrated on the bogie.

Hand-brake operable from the ground, on both sides of each bogie.

Axles and suspensions identical with the standard Y25 bogie.

Above each wheel is fixed a spark arresting shield.

These bogies are of a model regularly used for international freight operations. Line running tests have shown that the dynamic performance of a RoadRailer train equipped with these bogies is similar to train of conventional wagons equipped with the same bogies. The system is approved in the Continent and in the UK for 120 km/h operation.

3.5 Characteristics of the adaptors

The interface between the adaptor and the bogie, middle or end, is identical to the one on a conventional bogie wagon using a spherical centre pivot.

Every end adaptor is equipped with the set of buffing and draw gear required to couple with conventional railway equipment. The buffing and draw gear satisfies UIC requirements.

The end adaptors and the semi-trailers are designed to resist compression loads of 850 kN and tension loads of 1000 kN.

4. SAFETY DEVICES

4.1. At the coupling between semi-trailer and adaptor or between 2 semi-trailers

Three levels of safety is used to detect incorrect locking before the departure of the train.
- An audible device - a whistling is heard during the formation of the train.
- A visual device - flaps as well as the lid of the control box protrude on the underframe of the semi-trailer.
- An opening device of the main brake pipe stopping any movement of the train by preventing release of the train brake.

When correctly locked, the locking between the semi-trailer and the adaptor and between semi-trailers remains active due to mechanical devices (through redundant springs or gravity action). (Fig.2).

4.2. At the road axles level

Each axle is maintained and locked in the lifted position by two hooks.

Two devices allow the detection of an incorrect locking before the departure of the train.
- A colour marker.
- A device which allows the road axle to fall to the ground under gravity.

When the tractor unit is uncoupled from the semi-trailer during the build up of the train, the air pipe feeding the axle raising liftbags are depressurized, allowing the road axles to rest on their hooks or to drop to the ground. In the latter case, the non locking is immediately obvious.

Unlocking by the retraction of the two hooks can only be obtained when compressed air is supplied to the system (by coupling of the road tractor for example).

A pictogram, placed on the semi-trailers over the road axles, shows the upper position the axles must be in during rail operation.

4.3. At the trailer landing legs level

In addition to the irreversible mechanism of the landing legs, secured pins lock the legs as well as the handles in rail position.

A pointer allows the visual control of the correct retracted position of the landing legs.

A pictogram, placed on the semi-trailer above the landing legs, shows clearly this position.

5. CONCLUSIONS

The RoadRailer system has been evaluated by many of Europe's Railway Administrations. It has been approved by the UIC 5th Commission (as from the 1st July 1994) and vehicles have been registered with BR, DB and SNCF for both National and International commercial operations.

The first commercial service in Europe started June 1995 by BTZ, operating between Munich and Verona through the Alps with good reliability. The system operated without any incident under the harsh winter conditions. BTZ increased the original fleet of 60 curtainsider semi-trailers with 100 more trailers from which 60 are refrigerated vans, 20 tiltsiders and 20 dry vans. BTZ will be extending their operation to Northern Germany and Southern Italy.

In the UK RoadRailer is in commercial operation with Transrail, between Aberdeen and Northampton and soon will be establishing an Anglo-Italian service through the Channel Tunnel.

The French CNC start their commercial service within the next couple of months.

The benefits of the RoadRailer system: higher payload then any other form of combined transportation, lower traction costs due to the excellent aerodynamic characteristics, unified RIV UIC approved rolling stock, high average speed due to the good riding stability at high speeds (75-80 miles/h) and low handling costs will help rail to successfully compete with the road on the time-sensitive high value door to door transportation market.

RoadRailer have been operated for over a million km on European railways, without incident, showing the reliability of this innovative freight transportation system.

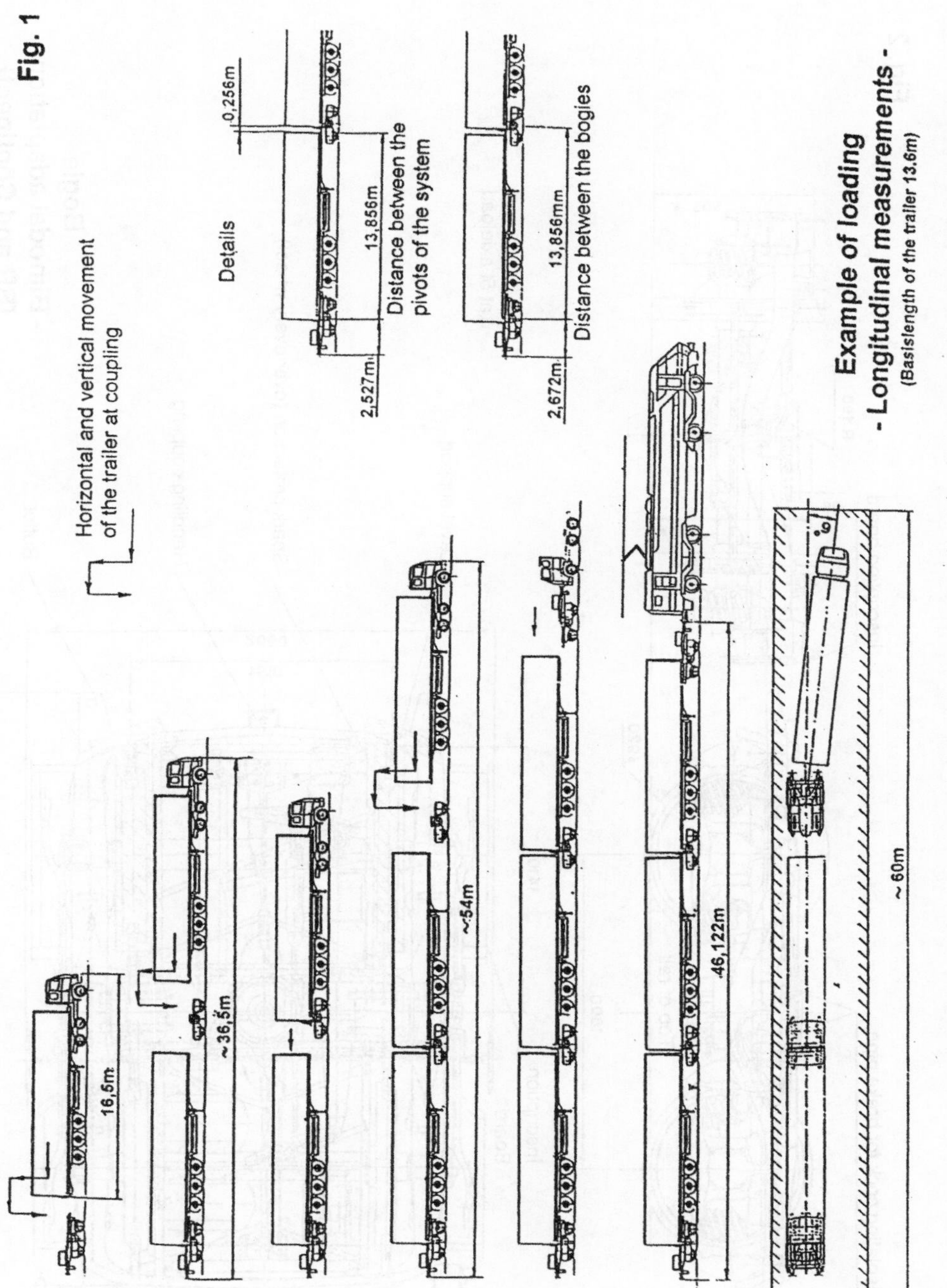

Fig. 1

Horizontal and vertical movement
of the trailer at coupling

Details

0,256m

2,527m

13,856m

Distance between the
pivots of the system

2,672m

13,856mm

Distance between the bogies

16,5m

~36,5m

~54m

46,122m

~60m

Example of loading
- Longitudinal measurements -

(Basislength of the trailer 13.6m)

Fig. 2

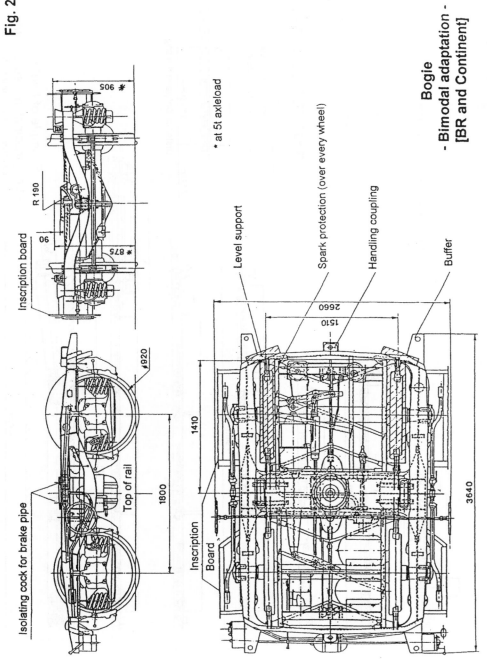

Inscription board

R 190

90

* 905

* 875

Ø 920

Isolating cock for brake pipe

Top of rail

1800

1410

Inscription Board

* at 5t axleload

Level support

Spark protection (over every wheel)

Handling coupling

Buffer

2660

1510

3640

**Bogie
- Bimodal adaptation -
[BR and Continent]**

Fig. 3

A-A

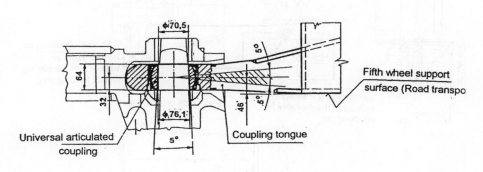

Universal articulated
coupling

Coupling tongue

Fifth wheel support
surface (Road transpo

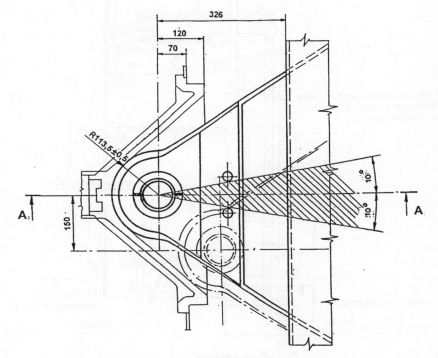

System coupling elements of trailer

C511/21/118 © IMechE 1996

133

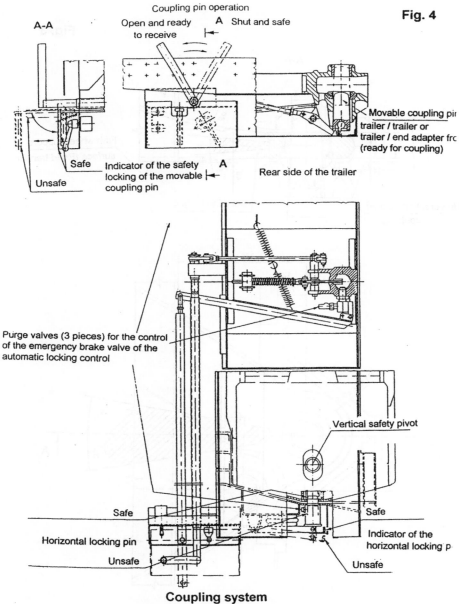

Fig. 4

Coupling pin operation

A-A

Open and ready to receive — A Shut and safe

Movable coupling pin trailer / trailer or trailer / end adapter frc (ready for coupling)

Safe

Unsafe

Safe — Indicator of the safety locking of the movable coupling pin — A

Rear side of the trailer

Purge valves (3 pieces) for the control of the emergency brake valve of the automatic locking control

Vertical safety pivot

Safe

Safe

Horizontal locking pin

Indicator of the horizontal locking p

Unsafe

Unsafe

Coupling system
- Construction characteristics -

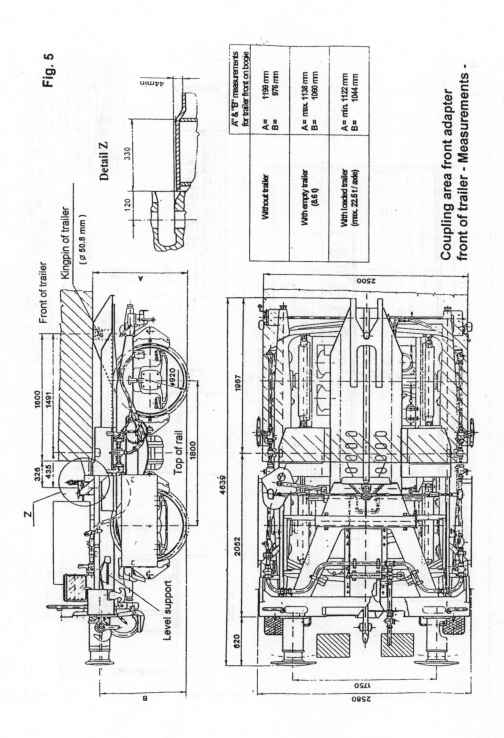

Fig. 5

Detail Z

44 min

330

120

Kingpin of trailer
(⌀ 50.8 mm)

Front of trailer

A" & "B" measurements for trailer front on bogie		
Without trailer	A=	1199 mm
	B=	975 mm
With empty trailer (8.6 t)	A= max. 1138 mm	
	B=	1069 mm
With loaded trailer (max. 22.5 t / axle)	A= min. 1122 mm	
	B=	1044 mm

Coupling area front adapter
front of trailer - Measurements -

⌀920

1600
1491

Top of rail

1800

328
435

Z

Level support

A

B

2500

1967

4639

2052

620

1750

2580

Fig. 6

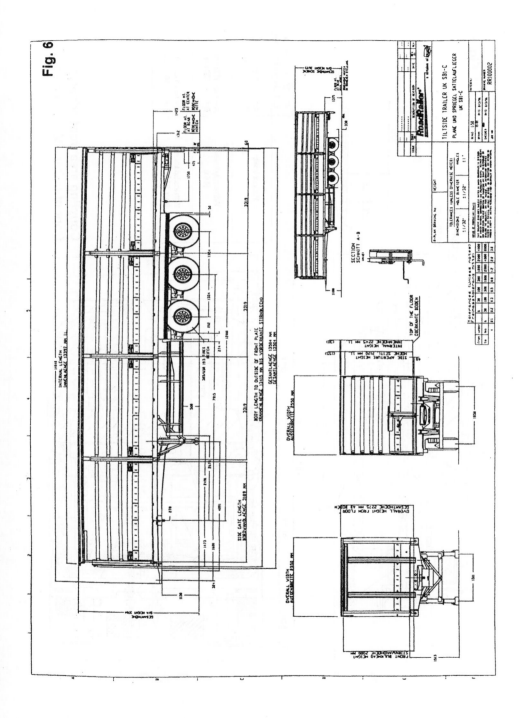

© IMechE 1996 C511/21/118

Fig. 7

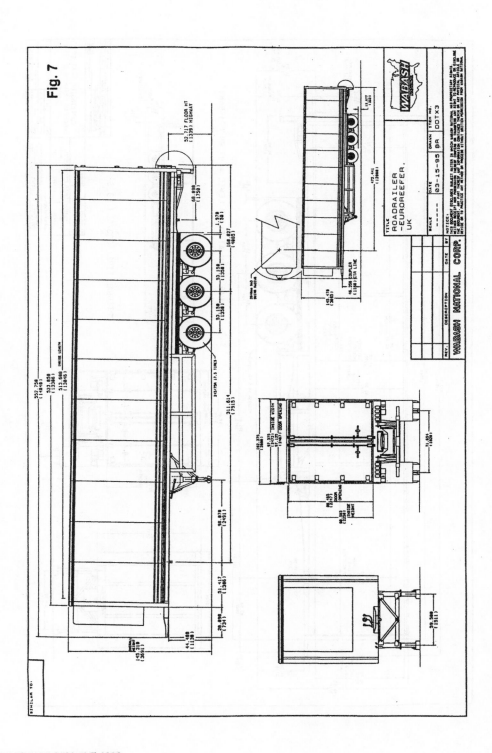

Fig. 8

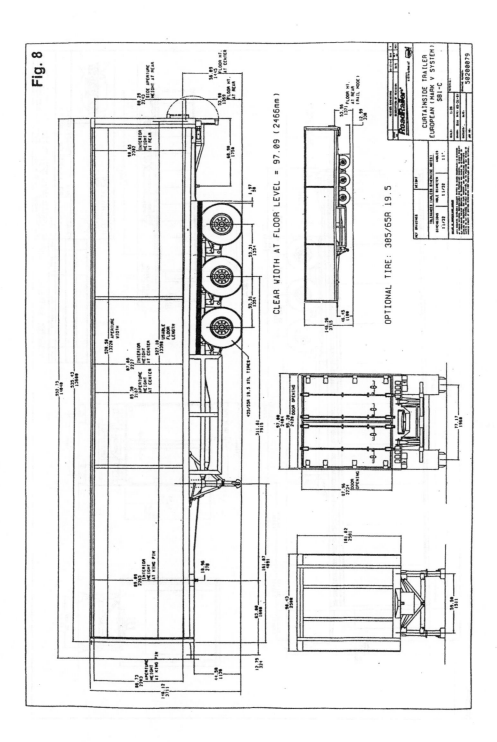

CLEAR WIDTH AT FLOOR LEVEL = 97.09 (2466mm)

OPTIONAL TIRE: 305/65R 19.5

Printed and bound by CPI Group (UK) Ltd, Croydon, CR0 4YY

16/04/2025

14658832-0001